Up from CLINICAL EPIDEMIOLOGY & EBM

O. S. Miettinen

Up from
CLINICAL EPIDEMIOLOGY
& EBM

Prof. Dr. O. S. Miettinen
McGill University
Cornell University
1020 Pine Avenue West
Montreal, QC H3A 1A2
Canada
olli.miettinen@mcgill.ca

ISBN 978-94-007-9764-2   ISBN 978-90-481-9501-5 (eBook)
DOI 10.1007/978-90-481-9501-5
Springer Dordrecht Heidelberg London New York

Printed on acid-free paper

Springer is part of Springer Science+Business Media (www.springer.com)

This text I dedicate to

Rebecca Fuhrer
my menschy leader
who had the inspiration to ask me to teach
a course on "clinical epidemiology"

and to

Johann Steurer
my highly learned colleague
who shares my zeal to ever better understand
clinical medicine and clinical research

as well as, of course, to

the young colleagues of mine
who are prescribed study of 'clinical epidemiology'
in preparation for practice of Evidence-Based Medicine

adding

Epidemiological Research, Terms and Concepts
as the called-for companion text
for simultaneous publication
by Springer

OSM
Montreal
November 2010

Acknowledgements

Rebecca Fuhrer, as the Chairperson of the Department of Epidemiology, Biostatistics and Occupational Health in the Faculty of Medicine of McGill University in Montreal, very kindly compensated for a major anomaly of the author of this text: complete agenesis of the skills of 'word processing' (in the modern meaning of this term) and the like. She unquestioningly saw to it that the needed help was available.

Kierla Ireland was the person actually providing the technical help. This she did with great competence and dedication, and in good cheer to boot.

Igor Karp volunteered as the Teaching Assistant (unpaid) for the course that gave rise to this text. Unsurprisingly, he functioned just like KI did. For he has been, for many years, a wonderfully dedicated student of the developments crystallized in this course text.

Johann Steurer had, unwittingly, a major role in the genesis of the contents of this text, in two ways. He was the inspiration for a good part of the relatively recent developmental work; and he supplied many of the recent books that substantially enriched the contents of this text (to say nothing about further educating its author).

Whatever may be the merits of this text, a good portion of these is due to the persons here gratefully acknowledged. It's been a privilege . . .

Foreword: On the Course Contents, Broadly

I feel honored that Professor Miettinen invited me to write a Foreword to this text, arising from his first-ever course on "clinical epidemiology." For I take this to be a text by the foremost thinker about clinical medicine and clinical research in today's era of Evidence-Based Medicine.

Miettinen first delineates the qualities of most productive leadership within the disciplines of clinical medicine and examines critically the present state of clinical academia. Then come the text's two principal sections, Theory of Clinical Medicine and Theory of Clinical Research, the former presented as the necessary foundation for the latter. A further section presents critical examinations of various EBM precepts and of a number of published studies.

Miettinen's propositions are to the effect that future clinical *academics* will be genuine experts on a given clinical topic insofar as they not only have had the requisite clinical experience but also are proficient in the theory of the relevant type of clinical research and have thoroughly reviewed the available evidence on the topic; that at present already, the tacit knowledge of clinical experts can, and should, be garnered into diagnostic and other expert systems; that future *practitioners'* professionalism will require their deference to the thus-codified expertise; and that teaching 'clinical epidemiology' to practitioners (of EBM) has been an anomalous response to the absence of such systems.

Some of the propositions the students may have been inclined to contradict. But, as they are results of Miettinen's very long-term critical and deep reflection on the topics, in conjunction with the abundance of his relevant erudition, those propositions should be discussed in the community of clinical academics and should challenge what now are mere received opinions in clinical academia.

J. Steurer, MD
Professor and Director, Horten Center for Patient-Oriented Research
and Knowledge Transfer
Department of Internal Medicine, Faculty of Medicine, University of Zurich,
Zurich, Switzerland

Preface: On Major Improvements in Clinical Medicine

"Over the past thousand years there has developed in the West a 'culture of improvement ... in which ... conditions have been cultivated to encourage and sustain improvement. Related to the value attached to improvement is the widespread expectation that improvement will indeed occur in most realms of technology."

Reference: Friedel R. A Culture of Improvement. Technology and the Western Millenium. Cambridge (MA): The MIT Press, 2007; p. 1.

While already highly technological, the vast industry of modern healthcare (ref. 1) nevertheless remains conspicuously bereft of the major improvements that *information technology* was expected to bring to it. The *knowledge-base of clinical medicine* still is not expressly, and in truly meaningful forms – much less comprehensively – codified in cyberspace, in diagnostic and other *expert systems*. At the very dawn of this Information Age, an eminent attempt was made to develop a diagnostic expert system; but it failed (ref. 2). The requisite theoretical understandings weren't yet there.

References:
1. Starr P. The Social Transformation of American Medicine. The Rise of a Sovereign Profession and the Making of a Vast Industry. New York: Basic Books, 1982.
2. Wolfram DA. An appraisal of INTERNIST – I . Artif Intell Med 1995; 7: 93–116.

Just recently, however, theoretical progress has produced understanding of the *forms* in which the knowledge-base of clinical medicine should be codified, and of the way its *content*, in terms of those forms, can and must be garnered from clinical experts' tacit knowledge. Thus the requirements now are in place for the development of practice-guiding expert systems, etiognostic and prognostic as well as diagnostic – for truly Information-Age practice of clinical medicine. By the same token, it now is clear what types of knowledge is to be pursued in clinical research to make those systems ever more scientific in their content.

Aims of the Course

The overarching aim of this course was to sow seeds of *major improvements* in clinical medicine in this Information Age (cf. Preface above).

One specific aim of this course thus was to orient some of the students – residents and fellows in the McGill University Health Centre – to the path through which they would become maximally *productive leaders*, and thereby agents of major improvements, in their respective disciplines ('specialties') of clinical medicine, now that the era of genuinely scientific medicine – its theoretical framework rational and its knowledge-base from science (ref. 1) – is dawning (ref. 2).

References:
1. Miettinen OS. The modern scientific physician: 2. Medical science versus scientific medicine. CMAJ 2001; 165: 591–2.
2. Miettinen OS, Bachmann LM, Steurer J. Towards scientific medicine: an information-age outlook. J Eval Clin Pract 2008; 14: 771–4.

Another specific aim of this course concerned all of the students. It was to cultivate in them resistance to the doctrines of the EBM (Evidence-Based Medicine) cult championed by the leaders of 'clinical epidemiology' and to orient them to the nature of rational, knowledge-based medicine, KBM, as well as to the R & D (research-and-development) that will serve to bring about and continually elevate universal excellence in KBM – thereby orienting the students to the path of becoming *genuine professionals* (ref.) in their respective disciplines of the practice of clinical medicine in this Information Age.

Reference: Miettinen OS, Flegel KM. Elementary concepts of medicine: X. Being a good doctor: professionalism. J Eval Clin Pract 2003; 9: 341–3.

Abstracts

Section I – 1

In clinical medicine, the role of its *leaders* is paramount. Academic leaders of the various disciplines of clinical medicine should take active interest in what amounts to genuine and well-qualified leadership in these disciplines, and in what should be understood to be their core responsibilities and missions. As for the latter, at this particular time, they should be in tune with the essence of modern public health and with the opportunities and challenges of this Information Age; and they should understand how study of 'clinical epidemiology' for the practice of Evidence-Based Medicine has been an anomalous response to the common absence of appropriate academic leadership in the disciplines of clinical medicine.

Section I – 2

A major mission in *medical academia* should be understood to be the teaching and advancement of the knowledge-base of scientific medicine. But the academia's success in this mission is seriously in question already on the superficial ground that there now are two, very different, conceptions of the essence of scientific medicine; more to the point, that both of these conceptions are profoundly wrongheaded – as, most notably, neither one of these conceptions involves any role at all for the knowledge-base of practice! This situation is due, in part, to the absence of the kind of critical discourse that in truly progressive academia would be seen to be essential. Fundamental re-orientation in the culture of medical academia is urgently needed.

Section I – 3

Had medical academia successfully defined scientific medicine, then, presumably, the clinical segment of this academia would have made a concerted effort to arrive at a consensus about the essence of the *clinical research*. But as it is, no such effort has been made, and the essence of clinical research remains a matter of very diverse opinions even among authors of textbooks on the subject; and the primacy of the research to advance the knowledge-base of clinical practice remains largely unrecognized. Remarkably, this seems not to bother the authors nor anyone else in clinical academia. A tenable conception of the essence of clinical research and of the most important genre of this is not only obviously important but also attainable.

Section I – 4

In today's medical academia, leaders of the clinical disciplines commonly refrain from voicing their views about the role of the established 'basic' sciences in the practice of medicine. Instead, they now commonly recommend that their junior colleagues *study 'clinical epidemiology'* as a new 'basic science' of clinical medicine. Yet, most of these leaders themselves have not studied its precepts on any level, let alone critically in the context of genuine expertise on clinical research. Only very critical study of these precepts actually is justifiable, and this even in terms that are very different between future practitioners and future academics.

Section I – 5

As the now commonly recommended study of 'clinical epidemiology' is meant to be preparation for practice of EBM, it is important, for everyone concerned, to understand that *EBM is a cult movement*, one whose founding doctrine is diametrically antithetical to the progress-promoting essence of modernity. The founding doctrine of the EBM cult is, also, at variance with the essence of science and the imperatives of professionalism in medicine. This doctrine is, however, widely appealing to doctors – even if its particular precepts, and practice, aren't. The prevailing discordance between the touting and the practice of EBM is huge. The proper balance between those two is attained by making both of them disappear.

Section II – 1

In medicine, as in general, a prerequisite for any thinking is possession of concepts; and correct thinking presupposes tenable concepts. So: Is the presence of M. tuberculosis a tenable conception of *the* cause of tuberculosis? Or is it, even, *a* cause of the disease? What is the essence of disease as a type of illness? Is Bayes' theorem germane to the theory of diagnosis? What, exactly, is diagnosis, and what is prognosis? What is correct in correct diagnosis and good in good prognosis? Is there 'gnosis' other than dia- and prognosis? Etc. Today's 'authoritative' dictionaries of medicine are not sources of the answers, tenable ones in particular. Nor is today's medical education (Apps. 1–2.) Unsurprisingly, thus, even the basic *essence of the knowledge-base of clinical medicine* remains generally ill-understood, even by clinical academics.

Section II – 2

The most profound conceptual challenge in contemporary medicine is that of grasping the *generic form of the ultimately relevant knowledge-base* of clinical medicine. It defines the terms in which the knowledge-base of clinical medicine now should be codified for expert systems in particular. As for diagnostic expert systems for primary care, is the form of the knowledge-base to be seen to be that of 'decision trees'?; or does it, instead, consist of likelihood ratios for the realisations of each of the diagnostic indicators that are based on the initial history-taking and physical examination in the context of a given type of patient presentation? Or, is there a conception more rational than either one of these? There is. And it is critically important to come to terms with this, including for fundamental orientation in the research to advance the knowledge-base of clinical medicine.

Section II – 3

A patient's presentation to a doctor would ideally lead to the same, maximally expert diagnosis, and prognosis (intervention-dependent), regardless of who the doctor is. This ideal can materialize if, and only if, the requisite knowledge of top experts on the topics has been suitably codified in cyberspace, for retrieval as needed in the course of practice. Toward this ideal, the first need is to understand *how expert clinicians' tacit knowledge, relevant to diagnosis and prognosis, can be garnered in the appropriate form.* The way this can be done has recently been described and illustrated. Thus, the theoretical basis for bringing about uniform, universally expert practice is now in place. It only needs to be translated into action, so that such practice gets to supersede the subjectivist dilettantism of EBM.

Section III – 1

Expert clinicians' diagnostic and other expertise – the ersatz knowledge this represents – derives, even at present, largely from their personal, extrascientific experiences with patient care. Their expertise is not particularly enhanced by evidence from such scientific experiences as now are being reported – whether concerning a diagnostic test's 'sensitivity' and 'specificity,' or the 'hazard ratio' as a characterization of an intervention's effect. Increasingly, however, experts' personal experiences now need to be supplemented by evidence from *scientific experience of the appropriate form*, thus allowing the expert systems to become the basis for increasingly scientific medicine.

Section III – 2

An expert clinician needs to take interest in the report on a piece of *diagnostic research* when its result is of the appropriate form – still very rare – and addresses the full complexity of a diagnostic challenge encountered in his/her particular discipline. When these requirements are satisfied, (s)he needs to determine how this result was produced, and on this basis to be able to evaluate the validity of its empirical content. And if (s)he deems the validity to be adequate, she needs to take interest in the precision of the diagnostic information in the result. For all this, (s)he needs true understanding of the theory of diagnostic research. This constitutes a challenge, notably as *fundamental fallacies* characterize today's 'clinical epidemiology' in this regard.

Section III – 3

Having achieved rule-in diagnosis, an expert clinician's next challenge may arise from the need to know about the causal origin – etiogenesis – of the patient's illness, possible iatrogenesis in particular. Different from diagnosis, the doctor's personal experience is not instructive about etiognostic probabilities; (s)he is totally depended on evidence from relevant research. But understanding the burden of that evidence from etiogenetic/*etiognostic research* is much more challenging than is its counterpart in respect to diagnostic research. In fact, it is so challenging that *fundamental fallacies* still characterize epidemiologists' research on the etiogenesis of illness, even though etiogenesis has been the principal concern in their research for a good half-century already.

Section III – 4

An expert clinician's prognostic knowledge can derive from his/her personal experience for relatively short-term prognoses only; and in these, as well as in long-term prognoses, modern medicine involves consideration of the effects of interventions as well as the intervention-conditional course of the illness at issue. *Prognostic research* is already very extensive, but it now generally is a matter of intervention trials with very *simplistic conception of the essential result*. But the data from these trials could be used to derive results of the appropriate form for the advancement of the knowledge-base of prognosis.

Section IV – 1

The leaders of the EBM movement act as authorities on how clinicians in general should critically evaluate the evidence from diagnostic and prognostic research. They issue precepts, and actual guidelines, for this evaluation. In these teachings they draw from epidemiology. But a point of major note is that diagnostic and prognostic research for clinical medicine have not been understood by epidemiologists any better than their central, etiologic/etiogenetic/etiognostic research for community medicine. Therefore, *the EBM precepts and guidelines on the evaluation of evidence are ill-founded*. They need to be taken with large grains of salt.

Section IV – 2

The leaders of the EBM movement are anything but profligate in illustrations of their guidelines for the evaluation of evidence, using contemporary – or whatever older – literature on clinical research. However, *examination of the nature of the now-prevailing culture in the production of scientific evidence* for clinical medicine is quite instructive. The students in this course had a major role in the selection of the example studies that were considered. Time and again their judgement was that the study, even if the result be taken at face value, is not truly relevant to their practices – on account of *inadequate form of the result* (reflecting inadequate design of the object of study).

Section V – 1

Looking back at this course, the point of departure in it was not commitment to 'clinical epidemiology' in preparation for practice of EBM. Instead, the commitment was to the Western *'culture of improvement,'* this with a view to healthcare in this Information Age. Now the knowledge-base of clinical medicine could, and should, be codified in *expert systems*. But it isn't. Promising for the future is, however, the fact that the form of the requisite knowledge is now understood, and also understood now is the way in which expert clinicians' tacit knowledge could be garnered in the form that is appropriate for those systems.

Section V – 2

The present situation confronts leaders of the various disciplines of clinical medicine with the challenge and topical mission to bring about major improvement – from status quo to *universal excellence* – in the practices. At issue is codification of the knowledge-base of the discipline in expert systems, thus providing for practice that is characterized by universal excellence in the *quasi-scientific* sense of this: the practice is, universally, like that of scientific medicine, except that its knowledge-base is derived from the tacit knowledge of experts with no inherent role for science as the source of this knowledge.

Section V – 3

The ultimate improvement for leaders of the disciplines of clinical medicine to help bring about is the transition from universally quasi-scientific medicine to universally *scientific* medicine. This requires cultivation of clinical research that truly is relevant for the advancement of the knowledge-base of practice in the discipline at issue – by virtue of the *form of the objects* of study. Contemporary clinical research remains seriously wanting in this critical respect, as was evinced by review of a number of eminent examples chosen by the students in this course.

Section V – 4

For bringing about the necessary major improvements in clinical medicine, ultimately providing for universal practice of scientific medicine, the *current teachings about 'clinical epidemiology' and EBM are counterproductive*, as they are not founded on tenable principles from the theory of medicine; and the EBM movement that they underpin is, philosophically, wrong-headed in its denial of the role of academic leaders and expert practitioners in the development of the knowledge-base of clinical medicine. The academic leaders, in turn, now consolidate this decadence by their promotion of 'clinical epidemiology' and EBM.

Appendices 1 – 2

In this course on 'clinical epidemiology,' the students – residents and fellows in various clinical disciplines – should already have gained secure command of general concepts of medicine, most elementary concepts in particular (cf. Sect. II – 1). To give them a sense of the extent to which they actually did have this background, they were given the assignment to define a particular set of elementary concepts of medicine. Their definitions were, generally, quite inconsistent (App. 1) and, thus, substantially at variance with what they should have been (App. 2).

Appendices 3 – 4

The students in this course were given a number of assignments specific to the content of the course (App. 3), to be addressed in group sessions of the students. The group experiences together with the teachers' subsequent handouts on how the assignments perhaps should have been dealt with (App. 4) were instructive: contrary to a central element in the founding doctrine of the EBM movement, gaining mastery of the theory relevant to true understanding of practice-relevant clinical research evidently commonly is too challenging for busy practitioners to take the time to acquire.

Appendix 5

Whereas a major theme of this course, antithetical to the founding doctrine of EBM, was Information-Age professionalism, with practice-guiding expert systems central to this, an important subordinate theme was the way in which clinical experts' tacit knowledge can be garnered into these systems. This introduces technicalities that did not belong in the course proper. Therefore, further orientation is given in this Appendix, specifically for the design of the set of hypotheticals to be presented to each expert panel.

Appendix 6

Whereas this course was, at its core, about major improvements that, I say, could and should be introduced into the vast industry of healthcare in this Information Age, and whereas I addressed these from my vantage of medical academia, I felt that all of this should be rounded out by a commentary from the perspective of industry at large. For this I needed someone with vast knowledge of industry at large and of matters scholarly pertaining to it, someone who also has a critical yet open mind together with commitment to the Western "culture of improvement" (Preface). So I asked my son to write this Appendix.

Contents

Acknowledgements   vii

Foreword: On the Course Contents, Broadly   ix

Preface: On Major Improvements in Clinical Medicine   xi

Aims of the Course   xiii

Abstracts   xv

PART I. PHILOSOPHICAL PROPEDEUTICS

I – 1.   ON LEADERSHIP IN CLINICAL MEDICINE   3

I – 2.   ON MEDICAL ACADEMIA AT PRESENT   5

I – 3.   PURPORTED ESSENCE OF CLINICAL RESEARCH   11

I – 4.   ON STUDY OF 'CLINICAL EPIDEMIOLOGY'   13

I – 5.   UP FROM 'CLINICAL EPIDEMIOLOGY' & EBM   15

PART II. THEORY OF CLINICAL MEDICINE

II – 1.   THE KNOWLEDGE-BASE OF CLINICAL MEDICINE:
ITS ESSENCE   23
On Concepts and Principles in General   23
The Essence of Clinical Medicine   24
Knowing about a Client's Health: Gnosis   26
The Knowledge-Base of Clinical Gnosis: Its Basic Essence   27
The Knowledge-Base of Clinical Gnosis: More on Its Essence . . .   28

II – 2.   THE KNOWLEDGE-BASE OF CLINICAL MEDICINE:
ITS NECESSARY FORMS   33
The Problem of Multiplicities   33
The Solution of the Multiplicities Problem: Functions   35
The Necessary Form of the Knowledge-Base of Diagnosis   37
The Necessary Form of the Knowledge-Base of Etiognosis   38
The Necessary Form of the Knowledge-Base of Prognosis   39

II – 3.   CODIFYING THE KNOWLEDGE-BASE OF EXPERT PRACTICE   41
          Knowledge-Base and Efficiency of Healthcare   41
          The Dream of Universal Excellence in Healthcare   42
          Requirements for Universal Excellence in Healthcare   44
          Meeting the Missing Requirement for Universal Expertise
          in Healthcare   45

PART III. THEORY OF CLINICAL RESEARCH

III – 1.   ENHANCEMENT OF PRACTICE BY CLINICAL RESEARCH .   53
           Evidence as the Product of Clinical Research   53
           Evidence as a Supplement to a Clinician's Experience   55
           Evidence in the Advancement of Clinical Knowledge   56
           Evidence in the Enhancement of Clinicians' Efficiency   57
           Priority-Setting for Quintessentially 'Applied' Clinical
           Research   57

III – 2.   INTRODUCTION INTO DIAGNOSTIC CLINICAL RESEARCH   59
           The Nature of the Results of Diagnostic Clinical Studies   59
           The Genesis of the Results of Diagnostic Clinical Studies   60
           The Quality of the Results of Diagnostic Clinical Studies   61
           Screening Studies as Exceptions in Diagnostic Clinical
           Research   63

III – 3.   INTRODUCTION INTO ETIOGNOSTIC CLINICAL RESEARCH   65
           The Nature of the Results of Etiognostic Clinical Studies   65
           The Genesis of the Results of Etiognostic Clinical Studies   66
           The Quality of the Results of Etiognostic Clinical Studies   67
           The 'Cohort' and 'Trohoc' Fallacies in Epidemiologists'
           Etiologic Studies   68

III – 4.   INTRODUCTION INTO PROGNOSTIC CLINICAL RESEARCH   71
           The Nature of the Results of Prognostic Clinical Studies   71
           The Genesis of the Results of Prognostic Clinical Studies   72
           The Quality of the Results of Prognostic Clinical Studies   74
           On Guidelines for Reporting on Clinical Trials   75

PART IV. CONTEMPORARY REALITIES IN CLINICAL RESEARCH

IV – 1.   ON EBM GUIDELINES FOR ASSESSMENT
          OF EVIDENCE   83
          EBM Precepts Overall: Their Assessments   83
          EBM Precepts re Diagnostic Research: Their Assessments   85
          EBM Precepts re Prognostic Research: Their Assessments   86

IV – 2.   SOME EXAMPLE STUDIES: THEIR ASSESSMENTS   89
          Examples in the Teachings about EBM   89
          Diagnostic Research: A Paradigmatic Study   93
          Diagnostic Research: Paradigm Lost, Example Series I   94
          Diagnostic Research: Paradigm Lost, Example Series II   97
          Etiognostic Research: Clinical Examples   99
          Prognostic Research: Clinical Examples   108
          Screening Research: Epidemiological Examples   119
          Screening Research: A Clinical Program   125

PART V.   EPILOGUE ON MAJOR IMPROVEMENTS
          IN CLINICAL MEDICINE

V – 1.   THE PREDICATES OF MAJOR IMPROVEMENTS   131

V – 2.   DEVELOPMENT OF THE MAJOR IMPROVEMENTS   133

V – 3.   RESEARCH FOR FURTHER IMPROVEMENTS   135

V – 4.   'CLINICAL EPIDEMIOLOGY' & EBM AS SET-BACKS   137

PART VI. APPENDICES

APPENDIX – 1.   SOME ELEMENTARY CONCEPTS
                OF MEDICINE, ACCORDING TO THE STUDENTS .   141

APPENDIX – 2.   ON THE STUDENTS' CONCEPTS,
                THE TEACHER'S COMMENTS   147

APPENDIX – 3.   ASSIGNMENTS TO THE STUDENTS   151

APPENDIX – 4.   TO THE ASSIGNMENTS, THE TEACHER'S
                RESPONSES   157

APPENDIX – 5.   MORE ON GARNERING EXPERTS' TACIT
                KNOWLEDGE   167

APPENDIX – 6.   AN INDUSTRIAL PERSPECTIVE   171

INDEX   175

PART I
PHILOSOPHICAL PROPEDEUTICS

I – 1. ON LEADERSHIP IN CLINICAL MEDICINE
I – 2. ON MEDICAL ACADEMIA AT PRESENT
I – 3. PURPORTED ESSENCE OF CLINICAL RESEARCH
I – 4. ON STUDY OF 'CLINICAL EPIDEMIOLOGY'
I – 5. UP FROM 'CLINICAL EPIDEMIOLOGY' & EBM

I – 1. ON LEADERSHIP
IN CLINICAL MEDICINE

Proposition I – 1.1: Whoever has become a *genuine* leader in whatever learned profession – one who actively sets the standards for thought and action by his/her followers, instead of merely perpetuating 'received opinions' and past practices – has been a follower, unwittingly perhaps, of Francis Bacon's precept, "Read not to contradict, nor to believe, but to weigh and consider" (ref. 1). Instead of having merely absorbed received opinions – or engaged in 'groupthink' (ref. 2) – (s)he has taken *personal responsibility for the tenets* (s)he has cultivated in his/her followers.

References:
1. Bacon F. The Essays or Counsels Civil and Moral. Oxford: Oxford University Press, 1999; p. 134.
2. Smolin L. The Trouble with Physics. The Rise of String Theory, the Fall of Science, and What Comes Next. Boston: Houghton Mifflin Company, 2006; pp. 286 ff.

Proposition I – 1.2: A genuine leader in a discipline ('specialty') of clinical medicine asks not how doctors think, as do, for example, Montgomery (ref. 1) and Groopman (ref. 2); (s)he asks how they *should* think.

References:
1. Montgomery K. How Doctors Think. Clinical Judgment and the Practice of Medicine. Oxford: Oxford University Press, 2006.
2. Groopman J. How Doctors Think. Boston: Houghton Mifflin Company, 2007.

Proposition I – 1.3: A *well-qualified* genuine leader in a discipline of clinical medicine eschews addled (muddled, confused) reasoning and, hence, submits to the 'mental legislation' (Kant) of the general *theory of medicine* (ref. 1) and its subordinate *theory of medical research*, specifically theory of quintessentially 'applied' clinical research – research of which (s)he, in his/her discipline, is a leader and a dedicated reviewer (ref. 2).

References:
1. Miettinen OS. The modern scientific physician: 7. Theory of medicine. CMAJ 2001; 165: 1327–8.
2. Miettinen OS. Evidence in medicine: invited commentary. CMAJ 1998; 158: 215–21.

O. S. Miettinen, *Up from CLINICAL EPIDEMIOLOGY & EBM*,
DOI 10.1007/978-90-481-9501-5_1, © Springer Science+Business Media B.V. 2011

Proposition I – 1.4: For a leader in a discipline of clinical medicine to bring about *major improvements*, (s)he is to satisfy two prerequisites beyond being genuine and well-qualified as a leader: like all contributors to the ascent of man (ref.), (s)he is to have "an immense integrity, and at least a little genius."

  Reference: Bronowski J. The Ascent of Man. Boston: Little, Brown and Company, 1973; pp. 144–8.

Proposition I – 1.5: In most industrialized countries, the greatest innovation in clinical medicine – and *public health* – since WW II has been this: making clinical medicine to join community medicine in the realm of public health – through the introduction of national health insurance. And major improvements in this vein are yet to come in respect, notably, to the central concerns in public-health policy, namely *quality assurance* and *cost containment* – with Information-Age R & D having a critical role in this (ref.). (The requisite nature of this R & D – the order in it actually is that of D & R – was the overarching object of this course.)

  Reference: Miettinen OS, Bachmann LM, Steurer J. Towards scientific medicine: an information-age outlook. J Eval Clin Pract., 2008; 14: 771–4.

Proposition I – 1.6: In the framework of Information-Age public health and its central concerns (propos. I – 1.5 above), an academic leader in a discipline of clinical medicine should see his/her first-order role to be to help bring about comprehensive, and ever better, codification of the discipline's *knowledge-base* – in a form suitable for practice-guiding expert systems. And while the thus-codified knowledge-base would be largely non-scientific in the years immediately ahead, (s)he should view his/her second-order role to be that of helping to make that knowledge-base ever more scientific – by cultivating the requisite original research (by junior colleagues) and actively engaging in (competent and) critical reviews of such research.

Proposition I – 1.7: The essential missions of a truly productive leader in a discipline of clinical medicine (propos. I – 1.6 above) are in sharp contrast with the anti-authority founding doctrine of Evidence-Based Medicine, put forward by leaders of 'clinical epidemiology' (ref.). Therefore, one important aspect of truly productive leadership in a discipline of clinical medicine now also is guiding one's followers to recognition of the fallacy in the idea that they should study 'clinical epidemiology' in prepareation for the practice of EBM.

  Reference: Evidence-Based Medicine Working Group. Evidence-based medicine. A new approach to teaching the practice of medicine. JAMA 1992; 268: 2420–5.

I – 2. ON MEDICAL ACADEMIA
AT PRESENT

There is only one justification for universities, as distinguished from trade schools.
They must be centers of criticism.

– Robert M. Hutchins

Proposition I – 2.1: "Professors pride themselves of objectivity, or failing that, fairness to competing views, or failing that, at least the capacity for neutral analysis. But ... Michèle Lamont (ref. 1) argues that professorial pride is excessive" (ref. 2). (Apropos, the young doctors taking this course were, and all readers of this course text are, encouraged to heed the precept in propos. I – 1.1 in respect to the teachings in this course.)

References:
1. Lamont M. How Professors Think. Inside the Curious World of Academic Judgment. Cambridge (MA): Harvard University Press, 2009.
2. Calhoun G, on the sleeve of Lamont's book.

Proposition I – 2.2: "Once we invest our opinion, we hang on to the investment; so the more we have at stake, the more we risk, even by doing nothing. And the more powerful we are, the more likely we are to stick to our rusty guns: because it was our firmness of purpose that made us powerful" (ref.).

Reference: James C. Cultural Amnesia. Notes in the Margin of My Time. New York: W. W. Norton & Company, Inc., 2008; pp. 507–8.

Proposition I – 2.3: "Twentieth-century medicine was struggling for the scientific footing that physics began to achieve in the seventeenth century. Its practitioners wielded the authority granted to healers throughout human history; they spoke specialized language and wore the mantle of professional schools and societies; but their knowledge was a pastiche of folk wisdom and quasi-scientific fads.... Authorities argued ... by employing a combination of personal experience, abstract reason, and aesthetic judgment." (It remains to be seen how the various authorities of 'clinical epidemiology' and EBM will argue in this 21st century, including in response to this course text.)

O. S. Miettinen, *Up from CLINICAL EPIDEMIOLOGY & EBM*,   5
DOI 10.1007/978-90-481-9501-5_2, © Springer Science+Business Media B.V. 2011

Reference: Gleick J. Genius. The Life and Science of Richard Feynman. New York: Pantheon Books, 1992; p. 132.

Proposition I – 2.4: In medical academia at present, a distinction is being made between *'basic'* and *'applied'* research. "The distinction [is viewed as one] between polite and rude learning, between the laudably useless and the vulgarly applied, the free and the intellectually compromised, the poetic and the mundane" (ref. 1). Accordingly, today's academics in clinical disciplines of medicine commonly seek to enhance their academic status by engaging in, or otherwise cultivating, 'basic' research; and regardless, they don 'basic' scientists' laboratory coats. An assertion of Barzun's may be apposite here: "When people accept futility and the absurd as normal, the culture is decadent" (ref. 2).

References:
1. Medawar P. Pluto's Republic. Oxford: Oxford University Press, 1982; p. 35.
2. Barzun J. From Dawn to Decadence. 500 Years of Western Cultural Life. 1500 to the Present. New York: Harper Collins Publishers, 2000; p. 11.

Proposition I – 2.5: Medical academia would do well adopting the view that all of medical research is 'applied' – application-oriented, in the meaning of having, by definition, the purpose of advancing the arts of medicine – and that its broadest subtypes are most meaningfully based on whether improved knowledge about the objects of study advances the *knowledge-base of medicine* (its practice, in the framework of already existing objects of practice-relevant knowledge). If the object of study is of this kind, the research is *quintessentially 'applied,'* the resulting knowledge being for application by practitioners. Otherwise the research is only *in-essence 'applied'* – potentially bringing something new to be addressed in quintessentially 'applied' medical research. The knowledge resulting from this deeper segment of medical research is of no professional concern to practitioners of scientific medicine. ('Applied' as a descriptor of research is less than apposite to denote its being motivated by application – potential or expected – of the knowledge being sought.)

Proposition I – 2.6: Whereas, per praxeologic theory, all of human action is aimed at advancing the actor's *personal happiness* (ref. 1), academic leaders – professors – of clinical disciplines of medicine should be persons who find personal happiness (ref. 2) in what actually is to be expected of them as agents of improvement: identification of deficiencies in the knowledge-base of their respective disciplines of (the practice of) clinical medicine and remedying these; that is, engagement in *purposive* – purpose-serving rather than interest-driven – clinical research and, specifically, in derivative – rather than original – clinical research of the quintessentially 'applied' sort (propos. I – 2.5 above; ref. 3). (The nature of this research, original and derivative, was a major object of this course; cf. propos. I – 1.5.)

References:
1. von Mises L. Human Action. A Treatise in Economics. New Haven: Yale University Press, 1963; pp. 3, 14.

2. Nettle D. Happiness. The Science Behind Your Smile. Oxford: Oxford University Press, 2005.
3. Miettinen OS. Evidence in medicine: invited commentary. CMAJ 1998; 158: 215–21.

Proposition I – 2.7: Some academic leaders in the disciplines of clinical medicine who have the appropriate disposition (propos. I – 2.6 above) presumably do not act accordingly; for something that Kant said in the context of polemics (ref.) presumably applies to some of today's academic leaders in clinical disciplines of medicine:

There is in human nature an unworthy propensity ... to conceal our real sentiments, and to give expression only to certain received opinions ... this conventionalism [constituting] the mischievous weed of fair appearances.

Reference: Kant I. Critique of Pure Reason (translated by Meiklejohn JMD). Amherst (NY): Prometheus Books, 1990; p. 420.

Proposition I – 2.8: The presumably furtively-questioned received opinions in medical academia in the previous century concerned, most importantly, the *essence of scientific medicine*. "By the end of the [1945–2000] period, [EBM] was advocated as the new approach and students [in the U.S. and U.K.] were taught to assess published accounts of treatment of patients, trial data of therapies, and the appraisal of relevant literature. This contrasted with the [Flexnerian] academic approach taught fifty years earlier, that clinical problems could be solved by the intellectual application of basic scientific principles" (ref. 1). "Investigation and practice are one in spirit, method and object," Flexner wrote (ref. 2); and his ideas actually remain well-respected by many in the medical academia of the present time, in competition with those underlying the EBM movement. Academics in each of the two camps conceal their real sentiments about the scientific-medicine concept of their colleagues in the other camp; and so, academic peace prevails in the framework of common conventionalism and maintenance of fair appearances (cf. propos. I – 2.7 above).

References:
1. Hardy A, Tansey EM. Medical enterprise and global response, 1945–2000. In: Bynum WF, Hardy A, Jacyna S, Tansey EM. The Western Medical Tradition. 1800 to 2000. Cambridge (U.K.): Cambridge University Press, 2006; p. 462.
2. Flexner A. Medical Education in the United States and Canada. Bulletin no. 4. New York: Carnegie Foundation for the Advancement of Teaching, 1910; p. 56.

Proposition I – 2.9: Given that medical academia has been and continues to be doctrinaire (and schizoid; propos. I – 2.8 above) yet seriously *mistaken* about the essence of scientific medicine – which truly is characterized by rationality of its theoretical framework and scientific origin of its (substantive) knowledge-base (ref. 1) – Kant (ref. 2) could have been describing the Flexnerian and EBM cultures of 20th-century medical academia when asserting that,

Where we find a complete system of illusions and fallacies, closely connected with each other and depending on grand general principles, there seems to be

required a peculiar negative code of mental legislation, which, under the denom-
ination of *discipline*, and founded upon the nature of reason and the objects of
its exercise ... shall constitute a system ... which no fallacy will be able to
withstand or escape from, under whatever guise or concealment it may lurk. ...
[For, as] reason is the source of all progress and improvement, [it] is to be held
sacred and inviolable.

References:
1. Miettinen OS. The modern scientific physician: 2. Medical science versus scientific
   medicine. CMAJ 2001; 165: 591–2.
2. Kant I. Critique of Pure Reason (translated by Meiklejohn JMD). Amherst (NY):
   Prometheus Books, 1990; pp. 399, 422.

Proposition I – 2.10: The needed 'mental legislation' is *theory of medicine*
(propos. I – 1.3) for a start; but, remarkably, medical academia at large still is devoid
even of the concept of this. And what should be understood to be its subordinate *theory of clinical research*, notably that of quintessentially 'applied' clinical
research (propos. I – 1.3, 2.5), is now uncritically being equated with 'clinical
epidemiology' (ref.).

Reference: Miettinen OS, Bachmann LM, Steurer J. Clinical research: up from 'clinical
epidemiology.' J Eval Clin Pract 2009; 15: 1208–13.

Proposition I – 2.11: A genuine future *leader* in a discipline of clinical medicine
weighs and considers the received opinions that now permeate clinical academia,
and to his/her dismay (s)he concludes that received opinions that are mere illusions
and fallacies are being perpetuated in the absence of the requisite mental legislation
founded upon the nature of reason. Rather than concealing his/her real sentiments
about all of this and pursuing fair appearances, (s)he sets out to find the path to
bringing about major improvements in his/her discipline of clinical medicine, understanding that submission to the dictates of reason is essential for success in this
noble mission. (S)he also understands, however, that bringing about major changes –
major improvements, even – won't be easy, as "human institutions tend to preserve
ideas like rock preserves fossils" (ref.).

Reference: Brown RH. Man and the Stars. Oxford: Oxford University Press, 1978; p. 171.

Proposition I – 2.12: All genuine future *professionals* in the disciplines of clinical
medicine, irrespective of whether they aspire to become leaders of their disciplines,
come to understand that at issue indeed are disciplines of clinical medicine, meaning
that genuine professionals in them function in conformity with the "mental legislation" (Kant) of the theory of medicine, and that in this rational theoretical framework
they deploy, to the maximal possible extent, the knowledge of top experts in their
respective disciplines. They come to understand that commitment to these principles constitutes the foundation for the development of genuine professionalism
in the practice of clinical medicine (ref. 1). Moreover, they come to see EBM as
being philosophically at variance with these principles and as being founded on

mere "illusions and fallacies" (Kant), and they come to understand that evidence from clinical research is not for consumption by practitioners (à la EBM) but for the advancement of the general knowledge-base of clinical practice (of knowledge-based medicine, KBM; ref. 2), of genuinely scientific medicine (cf. Aims of the Course).

References:
1. Miettinen OS, Flegel KM. Elementary concepts of medicine: X. Being a good doctor: professionalism. J Eval Clin Pract 2003; 9: 341–3.
2. Miettinen OS, Bachmann L, Steurer J. Towards scientific medicine: an information-age outlook. J Eval Clin Pract 2008; 14: 771–4.

Proposition I – 2.13: In truly *well-qualified* medical academia, each professor of a particular clinical discipline is not only an experienced clinician but also fully proficient in the theory of clinical medicine and of quintessentially 'applied' clinical research; and (s)he also is an expert on the current implications of such research in his/her discipline, this on account of his/her continual and maximally comprehensive (as well as fully competent) reviewing of the literature and (routine) discourse about this with his/her fellow professors of the discipline. Given such a clinical academia, the *knowledge-base* of any given discipline of clinical medicine can and must be taken to be that of its professorate, collectively across universities.

Proposition I – 2.14: In truly *well-functioning* medical academia, truly well-qualified professors of the clinical disciplines (propos. I – 2.13 above) implement a fundamental duality in the *education* they provide: they educate future academics – researchers and teachers – for the various disciplines of clinical medicine, and they educate-and-train future practitioners of those disciplines (with future clinical academics also undergoing this E & T). In the latter endeavor, the initial, entirely educational segment – quite short (2 yrs., say) – is the same for all students of medicine (incl. community medicine), constituting the true 'medical commons'; and the ensuing education-and-training is differentiated according to the students' particular disciplines of medicine, though with some overlaps between/among some of these. (Cf. study of engineering.)

Reference: Miettinen OS, Flegel KM. Medical curriculum and licensing: still in need of radical revision. Lancet 1993; 340: 956–7.

Proposition I – 2.15: Given understanding of the nature of medical academia that is not only truly well-qualified (propos. I – 2.13 above) but also truly well-functioning (propos. I – 2.14 above), a sad *conclusion* is ineluctable: the segment of medical academia that now embraces teaching of 'clinical epidemiology' and EBM – along with other subjects futile and absurd (propos. I – 2.4) – in the education of future practitioners of clinical medicine does not represent the ideal qualities of this august institution.

I – 3. PURPORTED ESSENCE OF CLINICAL RESEARCH

Proposition I – 3.1: Among the fallacies that now prevail in clinical academia are these varied and mutually conflictual, well less than scholarly conceptions of the essence of clinical research in textbooks on it:

1. "Foremost among [the clinical sciences] is clinical epidemiology ... the science of making predictions about individual patients by counting clinical events in groups of similar patients and using strong scientific methods to ensure that the predictions are accurate" (ref. 1).
2. "This book is about the science of doing clinical research in all of its forms: translational research, clinical trials, patient-oriented research, epidemiologic research, behavioral science and health services research" (ref. 2).
3. "[S]ome researchers have narrowly defined clinical research to refer to clinical trials ... while others have ... even include[d] animal studies, the results of which more or less directly apply to humans. ... I have chosen to adopt a 'middle of the road' definition ... research conducted with human subjects (or material of human origin) for which the investigator directly interacts with the human subjects at some point during the study" (ref. 3).
4. "I emphasize the evaluation of drugs throughout the book because drug testing is the dominant form of medical research ... " (ref. 4).
5. "The purpose of this book is to teach both the 'users' and 'doers' of quantitative clinical research. Principles and methods of clinical epidemiology are used to obtain quantitative evidence on diagnosis, etiology, and prognosis of disease and on effects of interventions" (ref. 5).

References:
1. Fletcher RH, Fletcher SN. Clinical Epidemiology. The Essentials. Fourth edition. Philadelphia: Lippincott Williams & Wilkins, 2005; pp. 2–3.
2. Hulley SB et alii. Designing Clinical Research. Third edition. Philadelphia: Lippincott Williams & Wilkins, 2007; p. xiii.
3. Glasser SP (Editor). Essentials of Clinical Research. Dortrecht: Springer, 2008; p. 4.

4. Gauch RA. It's Great! Oops, No It Isn't. Why Clinical Research Can't Guarantee the Right Medical Answers. Dordrecht: Springer, 2008; p. vii.
5. Grobbee DE, Hoes AW. Clinical Epidemiology. Principles, Methods, and Applications for Clinical Research. Boston: Jones and Bartlett Publishers, 2008; p. xi.

These 'definitions' are commented on, and a substitute definition is given, in Appendix 4.

I – 4. ON STUDY OF 'CLINICAL EPIDEMIOLOGY'

Proposition I – 4.1: Future clinicians are now commonly expected, by their preceptors, to study 'clinical epidemiology' as the 'basic science' that allows them to critically examine reports on clinical research in the practice of EBM (propos. I – 2.8; ref.). This expectation is, however, commonly advanced without even familiarity with, let alone competent critical examination of, the teachings under either 'clinical epidemiology' or EBM. It thus remains for the future clinician studying 'clinical epidemiology' to critically weigh and consider (propos. I – 1.1) the precept that calls for submission to those teachings, notably that precept's implicit predicate that study of 'clinical epidemiology' prepares him/her to better practice clinical medicine.

> *Reference*: Sackett DL, Haynes RB, Gyatt GH, Tugwell P. Clinical Epidemiology. A Basic Science for Clinical Medicine. Second edition. Boston: Little, Brown and Company, 1991.

Proposition I – 4.2: The main aim of a student's critical assessment of the purported need to study 'clinical epidemiology' naturally is to be classification of the precept as true or false. (This course, most notably through this section along with section IV – 1, is intended to help the student in this weighing and considering.) But it also may be worthwhile to entertain the psychological categories of bullshit (ref. 1) and humbug (ref. 2) for the precept. (A precept is bullshit if the preceptor is indifferent about whether it is true or false, humbug if it in itself is not a falsehood but in the context is intended to mislead.)

> *References*:
> 1. Frankfurt HG. On Bullshit. Princeton: Princeton University Press, 2005; pp. 33–4.
> 2. Black M. The Prevalence of Humbug. Ithaca: Cornell University Press, 1985; p. 143.

Proposition I – 4.3: Instead of any in-depth, uncritical study of 'clinical epidemiology,' the real need of medical students already, or failing this, of young doctors later, is to study the *theory of clinical medicine* and the *theory of clinical research*, the latter with focus on quintessentially 'applied' clinical research (cf. propos. I – 2.10) – with the needed depth in these studies greatly dependent on whether at issue is preparation for leadership – professorship – or some other *academic* position in a discipline of clinical medicine or, instead, professionalism in the *practice* of the discipline (propos. I – 2.14).

Proposition I – 4.4: Orientational, critical study of 'clinical epidemiology' and EBM, together with in-depth study of the theory of clinical medicine and the theory of clinical research, would be very well justifiable on the part of today's and future *academics* in clinical medicine (incl. today's professors; cf. propos. I – 2.13–15, 4.3 above). But as for today's and future *practitioners* of clinical medicine, all of those studies can well be replaced by coming to grips with the fact that practice even of scientific medicine is not science, and that practitioners thereby are not scientists (cf. propos. II – 1.8).

Proposition I – 4.5: A medical student or a young doctor contemplating study of 'clinical epidemiology,' or actually setting out to do this, would do well taking note of the fact that it has taken the instructor of this course half-a-century of concentrated post-medical-school effort to come to more-or-less secure understanding of, even, many of the elementary topics of the concepts and principles of clinical medicine and of their subordinate ones concerning directly practice-relevant clinical research.

Proposition I – 4.6: A medical student or a young doctor contemplating study of 'clinical epidemiology' in preparation for practicing EBM would do well pausing to think, even for a fleeting moment, how much effort would be involved in critical reading of all the relevant literature – whether a single reviewer would be able to cover it even on a full-time basis.

Proposition I – 4.7: A medical student or a young doctor contemplating study of 'clinical epidemiology' in preparation for practicing EBM would do well pausing to think, even for a fleeting moment, about the relative merits of (a) each practitioner in a given discipline of clinical medicine continually reviewing, with wanting competence, for themselves, a small part of the relevant literature, and (b) a set of experts continually reviewing (with full competence) practically all of the relevant literature – on behalf of, and for, all of the practitioners.

I – 5. UP FROM 'CLINICAL EPIDEMIOLOGY' & EBM

Proposition I – 5.1: *Epidemiology* is now 'officially' defined as "The study of the occurrence and distribution of health-related states or events in specified populations … and application of this knowledge to control health problems" (ref. 1). That addled thinking underlies this definition is particularly evident from the associated explication that "Study includes surveillance, observation, hypothesis testing, analytic research, and experiments." Needed actually are separate definitions for epidemiological research and epidemiology per se – the latter being (practice of) community medicine (ref. 2), and the former being implicit in this (per propos. I – 2.4–5; cf. App. 4).

References:
1. Porta M (Editor), Greenland S, Last JM (Associate Editors). A Dictionary of Epidemiology. A Handbook Sponsored by the I. E. A. Fifth edition. Oxford: Oxford University Press, 2008.
2. Miettinen OS. Important concepts in epidemiology. In: Olsen J, Saracci R, Trichopoulos D (Editors). Teaching Epidemiology. Third edition. Oxford: Oxford University Press, 2010.

Proposition I – 5.2: *Clinical epidemiology* is now 'officially' defined as "The application of epidemiological knowledge, reasoning, and methods to study clinical issues and improve clinical care," with the explication that "Research is conducted in clinical settings, is led by clinicians, and has patients as the subjects of study" (ref.). The confusion about the concept is well evident in the remarkable particulars of this 'definition' of 'clinical epidemiology' and those of its associated 'definitions' of clinical research (in propos. I – 3.1; App. 4).

Reference: Porta M (Editor), Greenland S, Last JM (Associate Editors). A Dictionary of Epidemiology. A Handbook Sponsored by the I. E. A. Fifth edition. Oxford: Oxford University Press, 2008.

Proposition I – 5.3: Medawar (ref. 1) denigrated "rhapsodic intellection" in research in general and called for deployment of "the humdrum process of ratiocination" in the spirit of Kant (i.a.). A case of mere rhapsodic intellection was the 'clinical-epidemiology' inspiration of D. L. Sackett, in which "it dawned on him that epidemiology and biostatistics could be made as relevant to clinical medicine as his research into the tubular transport of amino acids" (ref. 2). As a matter of plain

O. S. Miettinen, *Up from CLINICAL EPIDEMIOLOGY & EBM*,
DOI 10.1007/978-90-481-9501-5_5, © Springer Science+Business Media B.V. 2011

humdrum ratiocination, however, a *prerequisite* for understanding quintessentially 'applied' clinical research (propos. I – 2.5) is mastery of the elements of the theory of medicine (propos. I – 2.10, 4.3), ultimately as to the generic nature of the requisite knowledge-base of clinical medicine (ref. 3). In these terms, in this Information Age, understanding of clinical research is principally a matter of command of the *theory of the R & D* (in the sequence of D & R) leading to as-needed accessibility – through expert systems – of the entire, increasingly scientific knowledge-base of clinical medicine (cf. propos. I – 1.5). Making epidemiology and biostatistics into something they haven't been before is not involved in this.

References:
1. Medawar P. Pluto's Republic. Oxford: Oxford University Press, 1982; p. 1.
2. Sackett DL, Straus SE, Richardson WS, et alii. Evidence-Based Medicine. How to Practice and Teach EBM. Second edition. Edinburgh: Churchill Livingstone, 2000; p. ix.
3. Miettinen OS, Bachmann LM, Steurer J. Clinical research: up from 'clinical epidemiology.' J Eval Clin Pract 2009; 15: 1208–13.

Proposition I – 5.4: As background for critical understanding of 'clinical epidemiology' as the conduit to EBM, future *professionals* in the various disciplines of clinical medicine (propos. I – 2.12) do well taking note of the essence of *modernity*:

What is modernity, and even more its 'late' version, but the subjugation of subjectivity to objectivity, the personal to the methodically mechanical, the individual to the institutional, the contingent and the spontaneous to the rule of rule? (Ref. 1)

Today we are more than ever governed by rules that eliminate space for even the smallest exercises of judgment. These rules are created by both private and public authorities … all interested in minimizing the uncertainty associated with judgment. (Ref. 2)

What is peculiar to the modern world … is a narrative of human self-realization … The routines of disciplined work … are given a larger meaning through their place in the bigger story. Let's say I am a dedicated doctor, engineer, scientist, agronomer. My life is full of disciplined routines. But through these I am helping to build and sustain a civilization in which human well-being will be served as never before in history … The meaning of these routines, what makes them really worth while, lies in this bigger picture … (Ref. 3)

References:
1. Shapin S. The Scientific Life. A Moral History of a Late Modern Vocation. Chicago: The University of Chicago Press, 2008; p. 3.
2. Garsten B. Saving Persuasion. A Defence of Rhetoric and Judgment. Cambridge (MA): Harvard University Press, 2006; pp. 9–10.
3. Taylor H. A Secular Age. Cambridge (MA): The Belknap Press of Harvard University Press, 2007; p. 716.

Proposition I – 5.5: The very *antithesis* of these characterizations of modernity is the ('post-modern') *founding doctrine of the Evidence-Based Medicine cult,* formulated by the then – and current – doyens of 'clinical epidemiology':

> [The old] paradigm puts a high value on traditional scientific authority and adherence to standard approaches, and answers are frequently sought from direct contact with local experts or reference to writings of international experts. The new paradigm puts a much lower value on authority. The underlying belief is that physicians can gain the skills to make independent assessments of evidence and thus evaluate the credibility of opinions being offered by experts. It follows that clinicians should regularly consult the original literature ... in solving clinical problems and providing optimal patient care.
>
> *Reference*: Evidence-Based Medicine Working Group. Evidence-based medicine. A new approach to teaching the practice of medicine. JAMA 1992; 268: 2420–5.

Proposition I – 5.6: Rephrased, the founding doctrine of the EBM cult (propos. I – 5.5 above) is this: Clinicians at large can and should acquire competence in the assessment of research evidence on topics in their respective disciplines (through study of 'clinical epidemiology' and EBM); having gained the requisite competence, they can and should do this assessment quite comprehensively and continually on topics relevant to their respective disciplines; and having done this, too, to whatever extent, they should practice according to their own opinions on those topics, in disregard of the views of representatives, however eminent, of the respective scientific communities.

Proposition I – 5.7: "The underlying belief" of the purported new paradigm – EBM – obviously is tenable to the extent that many clinicians do have the aptitude for gaining competence ("the skills") to make independent assessments of evidence from clinical research. But to actually gain that competence, they have to devote the requisite effort to this end (beyond their background education in medicine at large and then education and training in a particular discipline of this) – years of full-time study, covering select, relevant topics in mathematics, probability theory, statistics, and philosophy of science; and, extensively, the theory of clinical medicine and the theory of quintessentially 'applied' clinical research (supplemented by study of the lingua franca of modern science – the English language – if need be).

Proposition I – 5.8: Doctors who do acquire this added education become *scientific experts* within their respective disciplines on the particular topics on which they (competently) review the entirety of the available evidence and suitably discuss it with their fellow experts on those topics. As scientific experts – members of the topic-specific scientific communities – they understand their role to be one of consensus-seeking in the context of the initially divergent opinions among the experts at large, respectful of the opinions of the others. As genuine experts they do not "evaluate the credibility [*sic*] of opinions being offered by [other] experts," while clinging to their own. This feature, among others, distinguishes (the select few of) genuine experts from (the masses of) such *conceited dilettantes* as are being promoted (propos. I – 5.5, 5.6 above) by the leaders of the EBM cult.

Proposition I – 5.9: The founding doctrine of the EBM movement raises an obvious question: How is it that something this *futile* and *absurd* can quickly get to be accepted – nominally at least – by many in clinical academia? Otherwise phrased, an obvious question concerns the now-common deference to the leaders of the EBM cult and its underlying 'clinical epidemiology,' and it is: What makes *intellectual decadence* like this possible in clinical academia? (Cf. propos. I – 2.4).

Proposition I – 5.10: Just like the Flexnerian notions that "Investigation and practice are one in spirit, method and object" and that clinical practice is "intellectual application of basic scientific principles" (propos. I – 2.8) – those principles purportedly learned by study of the 'basic' sciences of medicine in medical school – the founding doctrine of the EBM cult (propos. I – 5.5, 5.6 above) also is very appealing to many modern clinicians: "The medical profession has had an especially persuasive claim to authority. ... Its practitioners ... serve as intermediaries between science and private experience, interpreting personal troubles in the abstract language of scientific knowledge" (ref.). However, while appealing to the authority of science, doctors generally do not like to submit to the authority of scientific experts (propos. I – 5.5), nor do they really accept Claude Bernard's well-known precept, "Art is I, science is we." They don't like to see themselves as mere "intermediaries between science and [the client's] private experience"; they like to see themselves as actual scientists and, specifically, in the (grossly malformed) sense of 'Science is I.' Like those Flexnerian notions, the founding doctrine of EBM (propos. I – 5.5, 5.6) appears to have been designed to dovetail into these science-related anomalies in the self-image of many modern doctors.

Reference: Starr P. The Social Transformation of American Medicine. The Rise of a Sovereign Profession and the Making of a Vast Industry. New York: Basic Books, 1982; p. 4.

Proposition I – 5.11: Scientific knowledge is intersubjective (ref.); and therefore, anyone who in the practice of clinical medicine draws authority from science – as any practitioner indeed should, to the maximal realistic extent – should draw it from the *authority of the relevant community of scientific experts* on the state of scientific (evidence and) knowledge on the matter at hand (cf. propos. I – 5.8). For there can be no other genuine authority on a scientific matter. (Unfortunately, even this genuine authority on a scientific matter can be – and in matters relevant to medicine still quite commonly is – plain wrong [cf. propos. III – 1.7]. Where a practitioner presumes to know this to be the case, (s)he is to present his/her countervailing arguments to the client instead of simply ignoring the authority.)

Reference: Niiniluoto I. The nature of science. In: Niiniluoto I. Is Science Progressive? Dordrecht (NL): D. Reidel Publishing Company, 1984.

Proposition I – 5.12: Practitioners of clinical medicine should recognize as leaders of a mere cult – and thus as false leaders of thought in scientific medicine – those who put forward the founding doctrine of EBM (propos. I – 5.5, 5.6), on the basis of the nature of this doctrine alone. But more to the same effect is these leaders' arrogation to themselves a "growing ability to transform critical appraisals of evidence into direct clinical action" and asking of their followers "humility without which you will become immune both to self-improvement and to advances in medicine" – and "enthusiasm" as well as "irreverence" to boot (ref.).

Reference: Sackett DL, Straus SE, Richardson WS, et alii. Evidence-Based Medicine. How to Practice and Teach EBM. Second edition. Edinburgh: Churchill Livingstone, 2000; pp. ix, xii.

Proposition I – 5.13: Even truly well-qualified leaders of (particular disciplines of) clinical medicine (propos. I – 2.13) do not have the "ability to transform critical appraisals of evidence into direct clinical action" (cf. propos. I – 5.12 above); they do not presume to possess this translational prowess even in respect to available scientific knowledge. They understand that "Science never tells a man how he should act; it merely shows how a man must act if he wants to attain definite ends" (ref. 1). Otherwise put, they understand (with Lord May; ref. 2) that "The role of the scientist is not to determine which risks are worth taking, or deciding what choices we should take, but the scientist must be involved in indicating what the possible choices, constraints and possibilities are."

References:
1. von Mises L. Human Action. A Treatise on Economics. Third revised edition. Chicago: Contemporary Books, Inc., 1966; p. 10.
2. Pielke RA, Jr. The Honest Broker. Making Sense of Science in Policy and Politics. Cambridge (U.K.): Cambridge University Press, 2007; p. v.

Proposition I – 5.14: A practitioner of clinical medicine holds a professional position of public trust. (S)he therefore is obliged to measure up to what is expected of him/her as a professional: practice in deference to the leaders – top experts – of the discipline in whatever is the matter at hand (propos. I – 2.12, 5.4). (Any difficulty with this should lead to open critique of the leaders' ideas, possibly leading to a change in the 'guidelines' – norms – of practice). *Practice by the presumption of intrinsic superiority of one's personal opinions over those of experts on matters scientific – in the spirit of EBM* (propos. I – 5.5, 5.6) *– is antithetical to professionalism and betrayal of public trust.*

Proposition I – 5.15: With all this said about the EBM cult and its underlying nascent (and still inchoate) body of 'clinical-epidemiology' doctrines, and with the untenability of the latter made explicit in Part IV to follow, this question may nevertheless arise: Doesn't the rapid, wide acceptance of both 'clinical epidemiology' and EBM by many in clinical academia, first in their native Canada and then in a number of other countries, attest to tenability of their precepts? Arguably at least, the correct answer is: On the contrary. When Adam Smith had published his very enthusiastically received *Theory of Moral Sentiments*, David Hume reminded him that "Nothing, indeed, can be a stronger presumption of falsehood than the approbation of the multitude; and Phocion, you know, always suspected himself of some blunder when he was attended with the applause of the populace" (ref.). The multitude in medical academia that now expresses approbation of 'clinical epidemiology' and EBM is, generally, quite unfamiliar with the teachings under these two rubrics, much less has it engaged in critical (and competent) evaluation of these (propos. I – 4.1). By contrast, the populace of Athens in the fourth century BCE presumably was well familiar with, and also able to judge, the great statesman's public pronouncements.

Reference: Boorstin DJ. The Discoverers. A History of Man's Search to Know His World and Himself. New York: Vintage Books, 1983; p. 658.

Proposition I – 5.16: "Socrates' great merit is his probing, his making evident the flimsy basis on which 'opinions' were based and statements made" (ref.). He really would have had a field-day with the opinions and statements that now constitute the ideology of 'clinical epidemiology' and EBM.

Reference: O'Malley JW. Four Cultures of the West. Cambridge (MA): The Belknap Press of Harvard University Press, 2004; p. 78.

Proposition I – 5.17: "The attempt to push rational inquiry obstinately to its limits is bound often to fail, and then the dream of reason which motivates philosophical thinking seems merely a mirage. At other times, though, it succeeds magnificently, and the dream is revealed as a fruitful inspiration" (ref. 1). This course ultimately was about *the dream of reason* in clinical medicine (sect. II – 3). And, as "reason is the source of all progress and improvement" (propos. I – 2.9), this course was not about an idle dream; it was about expectation of major improvements that really could be brought about in clinical medicine (propos. I – 1.5). This dream really could succeed magnificently (sects. V – 1-3). For needed is, merely, submission to the dictates of reason; and "Though a generation is sometimes required to effect the change, scientific communities have again and again been converted to new paradigms" (ref. 2).

References:
1. Gottlieb A. The Dream of Reason. A History of Philosophy from the Greeks to the Renaissance. New York: W. W. Norton & Company, Inc., 2000; p. ix.
2. Kuhn TS. The Structure of Scientific Revolutions. Second edition, enlarged. Chicago: The University of Chicago Press, 1970; p. 152.

PART II
THEORY OF CLINICAL MEDICINE

II – 1. THE KNOWLEDGE-BASE OF CLINICAL MEDICINE: ITS ESSENCE
On Concepts and Principles in General
The Essence of Clinical Medicine
Knowing about a Client's Health: Gnosis
The Knowledge-Base of Clinical Gnosis: Its Basic Essence
The Knowledge-Base of Clinical Gnosis: More on Its Essence

II – 2. THE KNOWLEDGE-BASE OF CLINICAL MEDICINE:
ITS NECESSARY FORMS
The Problem of Multiplicities
The Solution of the Multiplicities Problem: Functions
The Necessary Form of the Knowledge-Base of Diagnosis
The Necessary Form of the Knowledge-Base of Etiognosis
The Necessary Form of the Knowledge-Base of Prognosis

II – 3. CODIFYING THE KNOWLEDGE-BASE OF EXPERT PRACTICE
Knowledge-Base and Efficiency of Healthcare
The Dream of Universal Excellence in Healthcare
Requirements for Universal Excellence in Healthcare
Meeting the Missing Requirement for Universal Expertise in Healthcare

II – 1. THE KNOWLEDGE-BASE OF CLINICAL MEDICINE: ITS ESSENCE

On Concepts and Principles in General

Proposition II – 1.1: To be able to *think at all* about matters clinical, a clinician needs *concepts* of the objects of clinical thought. A concept is the *essence* of a thing, that which is true of every instance of the thing – entity, quality/quantity, relation – and unique to it. A concept is specified by its definition. This posits the concept's proximate genus and its specific difference within this genus (as in: a triangle is a polygon with three sides). A term referring to a concept may consist of more than one word. (Some elementary concepts of medicine as they were defined by the young doctors taking this course are presented App. 1, and the teacher's comments on these are given in App. 2.)

> *Reference*: McCall RJ. Basic Logic. Second edition. New York: Barnes & Noble, Inc., 1952; pp. 1 ff.

Proposition II – 1.2: To be able to *think correctly* about matters clinical, a clinician needs *tenable* concepts – ones that are logically admissible (and in broader philo-sophical terms 'real') – of clinical medicine and, besides, the mental discipline of *principles* of clinical thought. Principles are 'synthetic' a-priori – solely reasoning-based – judgments/propositions. They govern thinking about clinical concepts and are, thus, 'augmentative' of those concepts (while 'analytic' a-priori propositions are deduced from concepts and are, thus, merely explicative of them).

> *Reference*: Kant I. Critique of Pure Reason (translated by Meiklejohn JMD). Amherst (NY): Prometheus Books, 2003; p. 7.

Proposition II – 1.3: The concepts and principles together with the requisite ter-minology of a given genre of human activity – games of chance, chess, sailing, tennis, musical composition, sample surveys, clinical medicine, clinical research, etc. – constitute the *theory* of that genre of activity. This conception of theory of human activity applies to those categories in which the generic types of challenge are essentially unchanging over time. Technology, notably, is not of this type. And so, till recently "There was not overall theory of technology – 'a coherent group of general propositions,' we can use to explain technology's behaviour" (ref.).

Reference: Arthur WB. The Nature of Technology. What It Is and How It Evolves. New York: Free Press, 2009; pp. 4, 23.

Proposition II – 1.4: With concepts and principles in general the concern in *logic* (within philosophy), Toulmin (ref.) distinguishes between "formal logic" and "material, or practical, or applied logic" or between "field invariant" and "field-dependent" logic. His discourse on logic is well worth a clinical scholar's attention. For in the theory of clinical medicine and the theory of clinical research, most of the logic – the propositions' basis in reasoning (as for the adoption of concepts and principles) – indeed is field-dependent, specific to clinical medicine and clinical research (as will become evident).

Reference: Toulmin SE. The Uses of Argument. Updated edition. Cambridge (U.K.): Cambridge University Press, 2003; pp. 172, 202.

The Essence of Clinical Medicine

Proposition II – 1.5: A *genuine clinical scholar, alas, needs to be quite critical about the concepts of medicine as they now are defined* – with mutual inconsistency (internally, even) – in "authoritative" dictionaries of medicine, starting from the concept of *medicine* itself (ref.):

– "the art of preventing or curing disease" (Stedman's);
– "the art and science of the diagnosis and treatment of disease and the maintenance of health" (Dorland's).

Reference: Miettinen OS, Flegel KM. Elementary concepts of medicine: I. Challenges with its concepts. J Eval Clin Pract 2003: 9: 307–9.

Proposition II – 1.6: The concerns in medicine are not about disease only, nor about disease and health only. Addressed in (the practice of) medicine is (a client's) 'health' in the inclusive meaning of this term, encompassing ill-health – *illness* – as well as freedom from illness, that is, health proper; and illness subsumes not only disease (L. *morbus*) but also defect (L. *vitium*) and injury (Gr. *trauma*). *Sickness* is overt manifestation of illness but can occur in health also (under behavioral or environmental stress).

References: Miettinen OS, Flegel KM. Elementary concepts of medicine: III. Illness: somatic anomaly with … ; IV. Sickness from illness and in health; V. Disease: one of the main subtypes of illness. J Eval Clin Pract 2003: 9: 315–23.

Proposition II – 1.7: Medicine is professional pursuit and attainment of *knowing* about a client's 'health' – more deeply or specifically than what is possible for laypersons – and *teaching* the client (or their guardian/representative) accordingly (L. *doctor*, 'teacher'). Intervention – whether preventive, therapeutic, palliative, or rehabilitative – is not in the essence of medicine. Even in modern medicine, despite the near-universal availability of modalities of intervention, a doctor rarely intervenes on the course of his/her client's health; it usually is the client who does the intervening, pharmaco-intervention, most notably. (A clinician's client in a given encounter is not inherently a *patient*, consulting the doctor because of suffering from sickness; L. *patient*, 'suffering.')

> *References*: Miettinen OS, Flegel KM. Elementary concepts of medicine: VIII. Knowing about a client's health: gnosis; IX. Acting on gnosis: doctoring, intervening. J Eval Clin Pract 2003: 9: 333–9.

Proposition II – 1.8: Medicine (as just defined) is *not science*. It is art, in the Aristotelian meaning of 'art' (ref. 1) – as in: the art of motorcycle maintenance (refs. 2–3). Nor is it any longer "the" art of anything. Instead, modern medicine at large is the aggregate of the differentiated arts/disciplines of medicine (ref. 4).

> *References*:
> 1. Miettinen OS. The modern scientific physician: 1. Can practice be science? CMAJ 2001; 165: 441–2.
> 2. Pirsig RM. Zen and the Art of Motorcycle Maintenance. New York: Bantam Books, 1979.
> 3. DiSanto RL, Steele TJ. Guidebook to Zen and the Art of Motorcycle Maintenance. New York: William Morrow and Company, Inc, 1990.
> 4. Weisz G. Divide and Conquer. A Comparative History of Medical Specialization. Oxford: Oxford University Press, 2006.

Proposition II – 1.9: Since no-one can master the entirety of modern medicine, there no longer is any true 'general practice' of medicine; there no longer are any true generalists among medical doctors. By the same token, there now also are *no specialists/specialties* either, only doctors/disciplines with particular definitions of what limited segment of medicine is their area of competence – general primary care, for example, characterized by breadth rather than depth of competence. (Cf. professional musicians, athletes, engineers, etc.: no one is said to be a specialist.)

Proposition II – 1.10: The aggregate of the disciplines of medicine at large is constituted by two first-order subaggregates, according to the broadest nature of the client: In *clinical* disciplines the doctor has multiple individual clients, cared for one at a time, while in the disciplines of *community* medicine – epidemiology (propos. I – 5.1) – the doctor has a single client, a particular population, being cared for as a whole or one subpopulation at a time (just as in clinical medicine the individual in a given instance may be cared for as a whole, or in respect to a particular segment of the whole).

Proposition II – 1.11: In respect to *education for quintessentially 'applied' research* (propos. I – 2.5), there now is a profound, though unjustifiable, difference between the academic cultures surrounding clinical and community medicine, respectively. Clinical academia at large (in schools/faculties of 'medicine') is, in all essence, devoid of the felt imperative to teach quintessentially 'applied' research to some of the students (cf. propos. I – 2.14), while in community-medicine academia (in schools of 'public health') such education – epidemiological – is as central a mission as any. In this anomalous situation, clinical researchers commonly have to seek ersatz investigator-education from community-medicine academia – in epidemiological research. (The teacher of this course did, among many others.)

Proposition II – 1.12: 'Academic medicine,' 'molecular medicine,' 'nuclear medicine,' 'experimental medicine,' etc., are *not* genuine disciplines of medicine; nor are various laboratory disciplines merely contributing to diagnosis (pathology, biochemistry, diagnostic radiology, etc.) or clinical disciplines of mere execution of intervention (surgical, radiological, . . .). To be sure, trauma surgery, for example, is a discipline of medicine, as defined (propos. II – 1.7), the specific difference (propos. II – 1.1) involving not only trauma as the object of the gnoses (propos. II – 1.13 below) and their consequent 'doctoring' (teaching, that is; propos. II – 1.7) but also the (potential) deployment, by the doctor, of surgery (as a genre of intervention; cf. propos. II – 1.7).

Knowing about a Client's Health: Gnosis

Proposition II – 1.13: In clinical medicine, a doctor's essential (i.e., definitional) work pertains to three fundamental subtypes of esoteric ad-hoc *knowing* (as a matter of its pursuit and attainment and, then, teaching the client accordingly; propos. II – 1.7):

– *dia*gnosis – knowing whether a particular illness was/is present;
– *etio*gnosis – knowing whether a particular antecedent (that was present, in lieu of its alternative) was causal – etiologic/etiogenetic – to the patient's illness (or mere sickness); and
– *pro*gnosis – knowing about the future course of the client's health in respect to a particular phenomenon of health (incl. how this course would depend on the choice of intervention).

These three constitute the first-order subtypes of the genus of esoteric ad-hoc medical knowing – of medical *gnosis*, that is.

Reference: Miettinen OS, Flegel KM. Elementary concepts of medicine: VIII. Knowing about a client's health: gnosis. J Eval Clin Pract 2003: 9: 333–5.

Proposition II – 1.14: In the pursuit of gnosis, the doctor ascertains a set of ad-hoc facts – (s)he needs to know what facts to ascertain – and then translates the set of facts – the gnostic *profile* – into the gnosis at issue – by bringing, for genuine gnosis, general medical knowledge to bear. The gnostic profile generally underdetermines the (particularistic, ad-hoc) truth of gnostic concern; and thus, gnosis generally can be probabilistic only. Perception of that (profile-specific, ad-hoc) probability (of a particular, potential truth about the health of a client) on the basis of general (abstract) knowledge is the *essence of medical gnosis* – genuine gnosis in medicine, that is.

The Knowledge-Base of Clinical Gnosis: Its Basic Essence

Proposition II – 1.15: In the translation of a diagnostic profile into the corresponding *diagnosis* about a particular illness, the doctor is to aim at attaining the corresponding *correct* diagnosis, characterized by the level of confidence/probability that the profile warrants (cf. propos. II – 1.14 above). Correct diagnosis is characterized by its level of probability being in accord (numerically) with the proportion of instances of the profile in general (in the abstract) such that the illness at issue is present: correct diagnosis about that illness is that probability of its presence (cf. propos. II – 1.14 above). (Diagnosis is not a guess, and correct diagnosis is not a guess that happens to be correct.) Thus the knowledge-base of diagnosis about a particular illness (in a particular instance, characterized by a particular diagnostic profile) is about the profile-specific prevalence of the illness in general (in the abstract). (When an illness is 'defined' as a mere syndrome of manifestations – rather than as their underlying somatic anomaly – diagnosis in this meaning of deeper, probabilistic knowing is replaced by mere pattern-recognition – as in the 'diagnosis' of, notably, 'mental illnesses' in general.)

> *Reference*: Miettinen OS. The modern scientific physician: 3. Scientific diagnosis. CMAJ 2001; 165: 781–2.

Proposition II – 1.16: In the translation of an etiognostic profile into the corresponding *etiognosis* about a particular antecedent (that was there), the doctor is to aim at attaining the corresponding *correct* etiognosis, characterized by the level of probability that the profile warrants. Correct etiognosis is characterized by its level of probability being in accord with the proportion of instances of the profile-cum-illness-and-antecedent in general such that the antecedent is causal – etiogenetic – to the case of the illness, that is, such that the antecedent completes a sufficient cause of the illness while its alternative would not (*ceteris paribus*): correct etiognosis about that antecedent is that probability of its etiogenetic role (in the case at issue). Thus the knowledge-base of a given etiognosis (about a particular antecedent in the context of a particular illness) is about the profile-specific etiologic/etiogenetic fraction (ref.) for the antecedent in general (conditionally on the antecedent having been there).

> *Reference*: Miettinen OS. Proportion of disease caused or prevented by a given exposure, trait or intervention. Am J Epidemiol 1974; 99: 325–32.

Proposition II – 1.17: In the translation of a prognostic profile into the corresponding *prognosis* about a particular (adverse) event of health in a particular range of prognostic/prospective time, or about a particular (adverse) state of health at a particular point in prognostic time, the doctor is to aim at attaining the corresponding *correct* prognosis, characterized by the level of probability that the profile together with a particular choice of intervention (preventive or therapeutic) warrants. Correct prognosis of this type is characterized by its level of probability being in accord with the proportion of instances of the profile-cum-intervention in general such that the event/state will occur in/at the particular period/point of prognostic time: correct prognosis about that event or state is that probability of its prospective occurrence. Thus the knowledge-base of such prognosis is about the incidence/prevalence – implying the probability – of the event/state at issue, specific for the prognostic profile, intervention, and period/point of prognostic time. The period/point of time that is of concern in prognosis may be that after/at the end of the course of a case of illness: prognosis may be about a particular outcome of the case of illness that is at issue (full recovery, a particular sequela, or fatality), and the knowledge-base of this type of prognosis is about the relative frequency of that outcome (conditionally on the course of the illness not being interrupted by death from another cause). (The concept of prognosis is to be distinguished from that of prediction/forecast, and thus the concept of correct prognosis is to be distinguished from correct prediction/forecast.)

The Knowledge-Base of Clinical Gnosis: More on Its Essence

Proposition II – 1.18: A *diagnostic profile* involves (in principle at least) two conceptually quite distinct subsets of diagnostic *indications*, realizations of diagnostic *indicators*. They constitute, respectively, the *risk profile* and the *manifestational profile*. The risk indicators are constitutional (congenital and/or acquired, commonly including age as an index of acquired constitutional characteristics), behavioral, and/or environmental; and the manifestational indicators are 'clinical' (based on 'history' – anamnestic and/or objective – and physical examination; L. *clinicus*, 'bed') in part at least, possibly supplemented by laboratory-based ones. Each of the indications is a realization on the indicator's *scale* – nominal, ordinal, or quantitative (difference or ratio) scale. Some of the indications may need to be characterized, also, in terms of their respective referents on a particular scale of *diagnostic time*, the zero point of which may be the time of the inception of the sickness prompting the pursuit of diagnosis.

Proposition II – 1.19: A *major misunderstanding* in the theory of diagnosis, still animating 'clinical epidemiology,' has been the idea (ref. 1) that general medical knowledge of the form of profile-specific prevalence/probability of the presence of a particular illness (propos. II – 1.15) is unrealistic to think/dream about; that the general knowledge-base of diagnosis (in any given instance) must be seen to be of the form of probabilities/likelihoods of the manifestational profile specific for each of the illnesses in the differential-diagnostic set; and that *Bayes' theorem* can be, and needs to be, used for the translation of the diagnostic profile into diagnostic probability on the basis of knowledge about those 'reverse probabilities' – for the translation of the probability prior to the manifestational facts into that posterior to the inclusion of these in the diagnostic profile.

References:
1. Ledley RS, Lusted LB. Reasoning foundations of medical diagnosis: symbolic logic, probability, and value theory aid our understanding how physicians reason. Science 1959; 130: 9–21.
2. Miettinen OS, Caro JJ. Foundations of medical diagnosis: what actually are the parameters involved in Bayes' theorem? Statist Med 1994: 13: 201–9.
3. Miettinen OS. The modern scientific physician: 3. Scientific diagnosis. CMAJ 2001; 165: 781–2.

Proposition II – 1.20: The Ledley-and-Lusted idea that certain theoretical disciplines "aid our understanding of how physicians reason" in the pursuit of diagnosis (ref. above) may be correct; but this idea should be understood to be irrelevant as a justification for that reasoning – as relevant in the theory of diagnosis is not how doctors reason but how they *should* reason (propos. I – 1.2).

Proposition II – 1.21: Different from the Ledley-and-Lusted idea that the profile-conditional probability/prevalence of an illness is prone to lack universality of value, it actually is universal so long as the diagnostic indicators are – and they always can be – formulated in universal terms (accounting, e.g., for the environmental level of the illness, endemic or epidemic, if relevant).

Proposition I – 1.22: Ledley and Lusted were mistaken also in their idea that illness-conditional probabilities of manifestational profiles are subject to general medical knowledge. These probabilities generally pose insurmountable epistemological challenges: valid study of these reverse probabilities requires assembly of cases of the illness at issue, and of its alternatives in the differential-diagnostic set, independently of the manifestational profiles – which generally is wholly impractical to accomplish. And even if valid assembly of the cases were feasible, the generally enormous number of different manifestational profiles in a given domain of presentation would make impossible the attainment of any semblance of reasonable precision for the probability estimates (apart from making the knowledge-base of diagnosis unmanageably complex).

Proposition II – 1.23: To that multiplicities problem 'clinical epidemiologists' have adduced a widely accepted false solution: replacement of the Ledley-and-Lusted idea (propos. II – 1.19) by that of sequential consideration of the component items in the diagnostic profile – thereby solving the epistemologic problem of multiplicities but introducing a serious ontologic one in its stead. The new problem is failure to account for the redundancies/intercorrelations among the indicators.

Proposition II – 1.24: *Screening* (for a cancer, notably) – pursuit of diagnosis before overt manifestation of the illness, to enable correspondingly early treatment – is a multifaceted topic that appropriately belongs in clinical, rather than community,

medicine. Yet, screening has up to now mainly been addressed by epidemiologists –
and as though screening were a matter of a single test and the application of it a
community-level preventive intervention. This continues to be yet another *major
misunderstanding* about diagnosis, one with particularly tragic consequences.

Reference: Miettinen OS. Screening for a cancer: a sad chapter in today's epidemiology. Eur
J Epidemiol 2008; 23: 647–53.

Proposition II – 1.25: In *etiognosis* the object of gnosis – presence/absence of
etiogenetic role for an antecedent that was there in lieu of its alternative – involves
two histories (the actual/factual one and its 'counterfactual' alternative) in respect to
a *risk factor*, a causal risk indicator, that is. Both of these histories generally should
be specified in respect to the entire range of potentially relevant *etiognostic time*,
the zero point of which is the time of the 'outcome' (inception/continuation of the
illness/sickness). Earlier history in respect to the risk factor at issue, even though
not etiogenetic, may need to be involved in the etiognostic profile (ref.), along with
other known risk factors, possibly supplemented by strong non-causal indicators of
the risk (e.g., age) – and, perhaps, also some specifics of the generic type of illness
(or sickness) at issue (e.g., cell type of lung cancer).

Reference: Miettinen OS, Caro JJ. Principles of nonexperimental assessment of excess risk,
with special reference to adverse drug reactions. J Clin Epidemiol 1989; 42: 325–31.

Proposition II – 1.26: A *major misunderstanding* of the essence of etiol-
ogy/etiogenesis of illness has been, and is, imbedded in this (ref.): "Of the numerous
changes that have occurred in medical thinking over the last two centuries, none
have been more consequential than the adoption of what Robert Koch called the
etiological standpoint" – thinking of, and (re)defining, diseases in terms of causes
that are universal, "common to every instance of a given disease."

Reference: Carter KC. The Rise of Causal Concepts of Disease. Case Histories. Burlington
(VT): Ashgate Publishing Company, 2003; pp. 1, 129 ff.

Proposition II – 1.27: When tuberculosis was redefined with involvement of a particular agent – *mycobacterium tuberculosis* – in the very concept of the disease,
as Koch famously did and others readily accepted, the disease actually got to be
redefined – but not renamed – as *mycobacteriosis* (cf., e.g., silicosis). But when the
presence of that agent was made intrinsic to the concept of the disease, this presence
ceased to be an antecedent to the inception of the disease, and it thus no longer could
logically be viewed as etiologic – causal – to the disease. Ditto for, say, HIV as 'the
cause' of AIDS.

Reference: Steurer J, Bachmann LM, Miettinen OS. Etiology in the taxonomy of illnesses.
Eur J Epidemiol 2006; 21: 85–9.

Proposition II – 1.28: The true proximal cause of a communicable disease, common to every instance of it, is not the agent involved; instead, it is 'effective
exposure' to the agent in conjunction with susceptibility to the exposure causing the

disease; and to be addressed for etiognosis in the context of a communicable disease thus are causes of these two inherently present proximal causes – their antecedents such as working with infected patients as to ergogenesis, and immunosuppressive therapy as to iatrogenesis.

Proposition II – 1.29: Different from the proportion/probability in the essence of correct diagnosis (propos. II – 1.15), the counterpart of this in etiognosis (propos. II – 1.16) is not subject to being known about on the basis of direct experience, not even in principle, as causation – causal connection – is not a phenomenon (perceived by the aid of the senses) but, instead, a *noumenon* (Kant's term for a "conception a priori"; ref. 1). For it to be studyable, etiognostic probability (P) needs to be thought of in terms of causally interpretable rate ratio (RR) contrasting the rate of the outcome's occurrence given the potentially etiogenetic antecedent against that given its alternative, this ratio specific to the etiognostic profile (p): $P = (RR_P - 1)/RR_P$ (ref. 2).

References:
1. Kant I. Critique of Pure Reason (translated by Meiklejohn JMD). Amherst (NY): Prometheus Books, 1990; pp. 2, 156 ff.
2. Miettinen OS. Proportion of disease caused or prevented by a given exposure, trait or intervention. Am J Epidemiol 1974; 99: 325–32.

Proposition II – 1.30: An empirical rate-ratio is causally interpretable (for etiognosis; propos. II – 1.29 above) if, and only if, it is descriptively valid and also free of *confounding* (refs.) by extraneous determinants of the outcome's occurrence – by being conditional on all potential confounders. (Various eminent sets of proffered 'criteria' for causality – Koch, Hill, Evans, . . . – are logically untenable.)

References:
1. Miettinen OS. Components of crude risk ratio. Am J Epidemiol 1972; 96: 168–72.
2. Miettinen OS. Confounding and effect-modification. Am J Epidemiol 1974; 100: 350–3.

Proposition II – 1.31: *Prognostic* profile is defined as of the zero point of prognostic time, as of the time prognosis is formulated. Thus the temporal referent of prognostic indicators is prognostic T_0. Prognosis is to be conditional not only on that profile but also on the intervention (preventive, therapeutic, or rehabilitative) that might or will be adopted (cf. propos. II – 1.13, 17). Prognosis in respect to intervention effect (conditional on the profile) – *intervention-prognosis* (causal) – is implicit in the difference between *descriptive prognosis* (acausal) conditional on the intervention and that conditional on its alternative. (All causal concepts involve a causal contrast; cf. propos. II – 1.16, 25.)

Proposition II – 1.32: While development of the scientific knowledge-base (acausal) of diagnosis has been held back by commitment to the theoretical framework of Bayes' theorem (propos. II – 1.19–23), that of the knowledge-base of prognosis has been retarded by commitment to the theoretical framework of *Cox regression*.

References:
1. Hanley J, Miettinen OS. Fitting smooth-in-time prognostic risk functions via logistic regression. Internat J Biostat 2009; 5: 1–23.
2. Miettinen OS. Etiologic study vis-à-vis intervention study. Eur J Epidemiology 2010; 25: 571–5.

II – 2. THE KNOWLEDGE-BASE OF CLINICAL MEDICINE: ITS NECESSARY FORMS

The Problem of Multiplicities

Proposition II – 2.1: The knowledge-base for *diagnosis* (within particular disciplines of clinical medicine) is to be organized by types of presentation for diagnosis, typically of the form of a given 'chief complaint' by a person from a particular (usually quite broad) demographic category. Within such a domain of presentation, the diagnostic indicators (propos. II – 1.18), even if rather few in number and simple in their scales, generally define (jointly) a *multiplicity* of possible diagnostic profiles and, thus, of subdomains for which the correct diagnosis – based on general prevalence (propos. II – 1.15) – is the object of needed general medical knowledge in respect to each of the illnesses the presence/absence of which is the object of diagnosis in that domain of diagnostic challenges. (Mere binary indicators, when only 10 in number, already imply $2^{10} = 1,024$ subdomains to be distinguished among.)

Proposition II – 2.2: For *etiognosis*, the knowledge-base is to be organized by type of illness (or sickness not due to illness) possibly together with the person's particular demographic category. Within such a domain, the etiognostic probability in respect to a given generic type of potentially etiogenetic antecedent (with a defined alternative; propos. II – 1.16) needs to be specific not only to a particular one of the possible etiognostic profiles (propos. II – 1.25) but, also, to a particular variant of that generic antecedent – commonly its levels in various segments of etiognostic time (propos. II – 1.25). The consequence is a *multiplicity* of situations for which knowledge about etiognostic probability – or about the causal rate-ratio that determines this (propos. II – 1.29) – is needed within any given domain of etiognostic challenges.

Proposition II – 2.3: For *prognosis*, the knowledge-base is to be organized by domains in which the role of a particular illness is of one of two kinds: either the illness already is present (according to rule-in diagnosis, based on a practically pathognomonic profile), or the illness (an overt case of it) is a futuristic concern (typically due to perceived, relatively high risk for it). These correspond to therapy-relevant and prevention-relevant prognosis, respectively; and rehabilitation-relevant prognosis also has its place here. Prevention-relevant prognosis already, regarding the possibilityof a particular illness emerging, generally involves a *multiplicity* of situations to be distinguished among, based on the prognostic (risk-related, subdomain-defining) profile and prospective risk factors (possibly including the choice of intervention); and in each of these the prognosis (in respect to the possible occurrence of the illness of concern) commonly needs to be specific to various periods/points of prognostic time (propos. II – 1.17). For a domain of therapy-relevant prognosis the multiplicity generally is even greater, on the basis of subdomains and interventions, for any given potential event/state among the phenomena characterizing the course of the illness or representing potential unintended effects of intervention.

Proposition II – 2.4: In general, for any realistic – suitably specific – codification of the knowledge-base of clinical medicine – of gnosis in it – there is a need to overcome *the problem of multiplicities* (propos. II – 2.1–3 above) – the problem that in any given domain of gnostic challenge (to know about a given object of gnosis) the knowledge is to be specific to a multiplicity of subdomains and possibly also to particulars of the object itself (as to its level and timing in etiognosis and timing in prognosis).

Proposition II – 2.5: Fundamental to *knowledge-based medicine*, KBM, scientific or not, is the principle that an instance of gnostic challenge from a given domain of presentation generally falls in a particular one of a multiplicity of operational (facts-based) categories, a subdomain of the domain of presentation that in the discipline is repeatedly encountered, and that it presents a need for correspondingly *specific general medical knowledge* (about frequency), existent or still nonexistent. *Rational medicine* is KBM with such distinction-making (in gnosis), as a matter of aspiration at least.

Proposition II – 2.6: The *anathema* of the fundamental principle of rational medicine (propos. II – 2.5 above; ref. 1) is expressed by the (to doctors quite nicely self-serving) adage – Kantian maxim (ref. 2) – that 'Every patient is unique and his own doctor knows best.' In this spirit, the EBM cult calls for "integrating the critical appraisal [of evidence] with our patient's unique biology, values and circumstances ..." (ref. 3). This passage, left without explication as it is, presents an insurmountable hermeneutical challenge (which is not uncommon in the precepts of the champions of EBM). From the vantage of reason, each patient encounter is unique; but for knowledge (general) to be relevant to it, it must be seen to be an instance of something general (abstract).

References:
1. Miettinen OS. Rationality in medicine. J Eval Clin Pract 2009; 15: 960–3.
2. Kant I. Critique of Practical Reason (translated by Abbott TK). Mineola (NY): Dover Publications, Inc., 2004; p. 17.
3. Sackett DL, Straus SE, Richardson NS, et alii. Evidence-Based Medicine. How to Practice and Teach EBM. Second edition. Edinburgh: Churchill Livingstone, 2000; p. 4.

The Solution of the Multiplicities Problem: Functions

Proposition II – 2.7: When, regarding the knowledge-base of clinical gnosis, the question is about the level of the *correct probability* for the gnosis at issue in a given domain for it, an orientational proper answer is: There is no single correct probability; it *depends*. Diagnostic probability for (the presence of) a given illness in a given domain of presentation depends on (the realizations for) the diagnostic indicators that (jointly) specify the subdomain at issue (propos. II – 1.15, 18). Etiognostic probability for (there having been an etiogenetic role for) a given antecedent (that was there) depends (jointly) on the etiognostic indicators that are being accounted for and the particulars of the (time course of the generic) antecedent (propos. II – 1.16, 25). And prognostic probability for (there being) a given event/state in/at a given period/point in the future course of the client's health depends (jointly) on the prognostic indicators and the choice of intervention (preventive, therapeutic, or rehabilitative; propos. II – 1.17, 31).

Proposition II – 2.8: The things on which the magnitude of something depends are in clinical jargon termed *determinants* of the magnitude. Thus, proposition II – 2.7 above can be recast in this form: The correct probability – in gnosis – depends (jointly) on the determinants of it that are accounted for (in making it suitably specific).

Proposition II – 2.9: Any well-understood way in which a probability/proportion – or any quantity – depends on its determinant(s) has to do with an expressly understood *domain* for this. Thus, if Y is a real-valued number that is determined by another number, X, in the sense of $Y = X^{1/2}$, the domain of this *function* inherently is that of $X \geq 0$, since for any $X < 0$ the square root is an imaginary number. The logarithm of a (real-valued) number, X, in the range $X < 0$ also is nonexistent, and hence the domain of log(X) also is $X \geq 0$. (Each of these two functions specifies an infinite number of values of Y, determined by the infinite number of the values of X.)

Proposition II – 2.10: When a quantity in a particular domain – category – of nature (in the abstract, as is the viewpoint of science) is thought of, or actually described, in terms of a particular, mathematical function of its determinant(s) within that domain, this function is termed a *model* for the relation at issue. Such a model is, as a matter of common definition, a formal, simplified representation of the relation at issue – inherently as to the *form* of the relation, but possibly also with *content* of that form (as to the magnitudes of the constants/parameters involved in the function). The function's form represents an adopted way to think about the relation (within the domain), and the possible content of that form represents either (some) experience per se or belief (subjective) or knowledge (intersubjective).

Proposition II – 2.11: *Knowledge* about gnosis-relevant probabilities specified by a model for a particular domain of clinical medicine is reasonably taken to be *experts' typical beliefs* about the magnitudes of those probabilities, philosophers' conception of knowledge notwithstanding. To the Platonic school of thought knowledge was, and in today's philosophy also commonly is, justified belief consistent with truth (ref. 1), even if some serious questioning of this has arisen (ref. 2). This conception of knowledge scarcely applies to empirical science, as evidence generally underdetermines the truth about the object of study (ref. 3). Indeed, scientific knowledge "can never be positively justified" (ref. 4); "All scientific knowledge is uncertain" (ref. 5).

References:
1. Boghossian B. Fear of Knowledge. Against Relativism and Constructivism. Oxford: Clarendon Press, 2006; p. 15.
2. Roberts RC, Wood WJ. Intellectual Virtues. An Essay in Regulative Epistemology. Oxford: Oxford University Press, 2007; pp. 5 ff.
3. Rosenberg A. Philosophy of Science. A Contemporary Introduction. Second edition. New York: Routledge, 2005; pp. 138 ff.
4. Popper K. Conjectures and Refutations. The Growth of Scientific Knowledge. London: Routledge & Kegan Paul, 2002; p. xi.
5. Feynman RP. The Meaning of It All. Thoughts of a Citizen-Scientist. Reading (MA): Perseus Books, 1998; p. 26.

Proposition II – 2.12: For the knowledge-base of clinical gnosis, the necessary (only practical) form – so long as the relevant distinctions (propos. II – 2.1–3) are being made – is that of *occurrence relations* (ref.), formulated as empirical models for the probabilities. For, focus on these *gnostic probability functions*, GPFs, commonly reduces the need to know, separately, about an enormous multiplicity (thousands) of probabilities for a given object of gnosis in a given domain, to the need to know about the magnitudes of the very much smaller number (at most dozens) of parameters involved in a reasonable model that addresses all of those probabilities.

Reference: Miettinen OS. Knowledge base of scientific gnosis. J Eval Clin Pract 2004; 10: 353–67.

Proposition II – 2.13: The complete set of GPFs for a particular domain of gnosis constitutes the *entirety of the knowledge-base* of gnostic practice concerning presentations from the domain. For, that set, as a complete set for the domain, implies, for one, the set of objects for gnosis in the domain – the differential-diagnostic set for a domain of presentation with a complaint (for diagnosis) as for the presence/absence of each of these; the set of potentially etiogenetic antecedents to consider for the domain of a case of a particular illness/sickness (for etiognosis) as for the etiogenetic role of each of these; and the prospective events/states to consider (for prognosis) in a given domain as for the occurrence/non-occurrence of each of these. For another, the complete set of GPFs for a given domain implies the complete set of both interventions to consider (for prognosis) and the gnostic indicators to be accounted for in the initial gnostic profile (for any gnosis), and also those in possible expansions of the profile (for diagnosis in particular); and this set of GPFs gives what ultimately is needed: the gnostic probabilities for each of the possibilities.

Reference: Miettinen OS, Bachmann LM, Steurer J. Clinical research: up from 'clinical epidemiology.' J Eval Clin Pract 2009; 15: 1208–13.

The Necessary Form of the Knowledge-Base of Diagnosis

Proposition II – 2.14: For *diagnosis* – for the probability of the presence of a particular illness in an instance from a particular domain – the particular generic form of the GPF generally is the *logistic* one:

$$\log[P/(1 - P)] = B_0 + \Sigma_i B_i X_i = L, \; P = 1/[1 + \exp(-L)],$$

where P is the diagnostic probability; $\log[P/(1 - P)]$ is the 'logit' (transform/metameter) of P; the Xs $(X_1, X_2, \ldots)$ are statistical variates (numerical) adopted ad hoc to represent the diagnostic indicators; the Bs $(B_0, B_1, B_2, \ldots)$ are the set of parameters that constitutes the object of diagnosis-relevant general knowledge regarding the illness at issue in the domain at issue in terms of this form for the knowledge; Σ_i stands for 'summation over i' $(i = 1, 2, \ldots)$; and 'exp' stands for 'exponential of,' meaning 'antilog base e of.'

Proposition II – 2.15: With the logit of P as the explanandum, the explanans $B_0 + \Sigma_i B_i X_i$ is *linear* in the meaning of being a 'linear compound' of the parameters $\{B_i\}$ (of which the value of the logit of P is composed, with the Xs the coefficients in the linear compound of the Bs, incl. $X_0 = 1$).

Proposition II – 2.16: That the model is linear for the logit of P – and, hence, nonlinear, or 'generalized linear,' for P itself (cf. propos. II – 2.14) – in no way restricts the forms of the relations that can be addressed in its framework. For example, for the logit of P as a quadratic function of age, considered alone, the linear compound is $B_0 + B_1 X_1 + B_2 X_2$ with X_1 representing (the numerical value of) the age per se and $X_2 = X_1^2$.

Proposition II – 2.17: For the knowledge-base of *decisions* about the use of a (set of) diagnostic *test*(s) – in a 'decision node' with its associated diagnostic profiles based on a particular set of diagnostic indicators – needed is, first, the relevant functions for *pre-test* probabilities regarding each of the illnesses of practical – sufficiently high-priority – concern in the differential-diagnostic set; and second, the *post-test* counterparts of these functions. For a given one of the illnesses, the corresponding pre-test function specifies the pre-test probability (by its realization at the pre-test profile); and when this is not extreme enough, the corresponding post-test function allows identification of the range of possible post-test probabilities (by its realizations at the pre-test profile supplemented by the positive and negative extrema of the test result[s]).

Proposition II – 2.18: For the purpose of decisions about the use of a (set of) test(s), an additional need beyond knowing the range of possible post-test probabilities (propos. II – 2.17 above) is to know about the probability of the testing leading to a practical rule-in, or rule-out, post-test diagnosis about a particular illness in the differential-diagnostic set. For the decision about a single test, the post-test function (evaluated at the pre-test profile) implies the range, if any, of test-result values that imply a practically 'conclusive' post-test probability in a given one of the two directions (practical rule-in and/or rule-out diagnosis), given that the corresponding extremum has this implication. Thus, for a situation in which a 'conclusive' result – scalar – of a given test (in a given direction) is possible, needed is, specifically, a function for setting the probability for this range (defined ad hoc) – a logistic function in which the probability of this range of the test result replaces the probability of the illness being present. When the decision concerns a test with vector-valued result or a set of tests, the counterpart of the unidimensional result range for a single test can be taken to be the sum of the terms in the post-test model that pertain to the component results in the set.

The Necessary Form of the Knowledge-Base of Etiognosis

Proposition II – 2.19: As the knowledge-base of etiognosis in the context of a given outcome fundamentally is about causal rate-ratios (implying etiognostic probabilities; propos. II – 1.29), that knowledge-base is constituted by functions each of which expresses the *outcome's rate of occurrence* in relation to a particular one of its various known etiogenetic determinants in a defined domain, with the relation made *causally interpretable* by its conditionality on potential confounders through suitable representation of these co-determinants in the rate function. (The outcomes of concern in respect to iatrogenesis commonly are matters of sickness without illness.)

Proposition II – 2.20: Akin to diagnostic probability functions, the rate functions for etiognosis are generally – and properly – given a 'generalized linear' form; that is, a linear (in the parameters) form is given to a suitable transform of the rate (R):

$$f(R) = B_0 + \Sigma_i B_i X_i = L; R = f^{-1}(L).$$

For a proportion-type rate (of incidence or prevalence) the generally appropriate transformation is the logistic one: $f(R) = \log[R / (1 - R)]$. For incidence density – number of events (expected) per unit amount of population-time – it is $\log(R')$, where R' is the numerical value of the rate.

Proposition II – 2.21: Given such a formulation for the rate function, the corresponding function for *rate ratio* is of the form

$$RR = f^{-1}(L)/f^{-1}(L_0),$$

where L_0 is L evaluated at the reference category (or level) of the etiogenetic determinant of the rate's magnitude, representing the alternative to the causal category (or level).

Proposition II – 2.22: Specifically, for a proportion-type rate the causal rate-ratio function in the context of a logistic model for the rate itself is

$$RR = [1 + \exp(-L_0)]/[1 + \exp(-L)],$$

while in the context of a log-linear model for incidence density it is

$$RR = \exp(L - L_0).$$

Proposition II – 2.23: While, for etiognosis, the domain generally is principally defined by the presence of the illness whose etiogenesis is at issue (possibly supplemented by some demographic and/or other characteristics), for the rate functions relevant for etiognosis the *domains* naturally are defined without any reference to the illness per se but only, possibly, to some determinant(s) of its rate of occurrence.

The Necessary Form of the Knowledge-Base of Prognosis

Proposition II – 2.24: In prognostic functions, an important distinction is that between *acute* – very short-term – prognosis, in which only the outcome of an already existing illness (at the end of the very short prognostic time period) or the occurrence of a complication of the illness or its treatment anywhere in that short period really matters, and *subacute or chronic* prognosis, in which subintervals of the prognostic time period and/or particular points in this period need to be considered (intervals for events, points for states; cf. propos. II – 1.17).

Proposition II – 2.25: For *acute* prognosis, a prognostic function is to address the proportion of instances of the various subdomains of the prognostic domain such that the outcome at issue (fatality, particular sequela, or full recovery from the illness), complication, or adverse event possibly due to intervention (propos. II – 2.3) occurs. Accordingly, the appropriate generic form of the function for acute prognosis generally is the logistic one.

Proposition II – 2.26: For *subacute or chronic* prognosis, a distinction is to be made according as at issue is a possible future state (of health) or, instead, a possible future event (e.g., the inception of a state). When at issue is a *state*, the appropriate function addresses, again for various subdomains of the domain of prognosis, prevalence/proportion/probability and is, thus, logistic in form; but different from a function for acute prognosis, prognostic time is to be included as a determinant of the prognostic probability (jointly with the prognostic indicators and intervention).

Proposition II – 2.27: For *subacute or chronic* prognosis about an *event*, the model is to address the event's incidence density, for (the numerical value of) which a log-linear formulation – $\log(ID) = L$; $ID = \exp(L)$ – generally is appropriate; and, of course, prognostic time again is to be included among this rate's determinants (together with the prognostic indicators and intervention). The model for ID implies the corresponding function for cumulative incidence, CI, and, thereby, for the prognostic probability, P, of the event's occurrence in the prognostic time interval from t_1 to t_2, as follows:

$$CI_{t_1,t_2} = P_{t_1,t_2} = 1 - \exp[- \int_{t_1}^{t_2} ID_t \, dt].$$

Reference: Miettinen OS. Estimability and estimation in case-referent studies. Am J Epidemiol 1976; 103: 30–6.

Proposition II – 2.28: As probability functions for subacute or chronic prognosis are to provide for addressing prognostic probabilities (for particular phenomena of health for particular periods/points of prognostic time) conditionally on (prospective) intervention in addition to the prognostic profile (at prognostic T_0), the question arises whether the prognoses should be *conditional on otherwise surviving*, on being there in the future to potentially experience the state or event at issue. To be generally meaningful they should, and the formulation in proposition II – 2.27, above, inherently involves this conditionality.

Proposition II – 2.29: The *domain* for a prognostic incidence density or probability function is to be one based on presence/absence of a particular illness (propos. II – 2.3) and, broadly, on one or more of the prognostic indicators (at prognostic T_0). These specifications generally imply presence of an indication for the interventions to consider and absence of contra-indications for these. For the domain, the *function* involves variates representing the prognostic indicators together with the (type of prospective) intervention, and for subacute or chronic prognosis prognostic time besides.

II – 3. CODIFYING THE KNOWLEDGE-BASE OF EXPERT PRACTICE

Knowledge-Base and Efficiency of Healthcare

Proposition II – 3.1: A. L. Cochrane, in a famous and very influential booklet (ref.), concerned himself with the National Health Service of the U.K. in respect to a way to enhance its effectiveness (in preserving and restoring health) in relation to its cost – enhancement of its efficiency in this meaning. His premise was that if doctors were able to know which one among the available options for intervention on any given indication is most effective, these interventions would be more commonly used (in lieu of less effective ones) and the effectiveness of the NHS would thereby improve. From this he deduced the need to cultivate *clinical trials* to assess the relative/comparative effectiveness of the available options for intervention. (The extent to which clinical trials themselves – generally quite expensive, requiring replications and ultimately 'Cochrane reviews' – actually have been cost-effective in enhancing the efficiency of the NHS of the U.K. or of other systems of healthcare remains unclear, however.)

Reference: Cochrane AL. Effectiveness and Efficiency. Random Reflections on Health Services. London: Nuffield Provincial Hospitals Trust, 1972.

Proposition II – 3.2: A modification of Cochrane's premise (above) deserves consideration: If doctors were able to know, right in the course of their practices, in respect to the type of situation that confronts them at a given moment, what their most illustrious colleagues in the same situation typically do (as a matter of fact-finding) or think (as a matter of translating the available facts into the corresponding gnosis), they would tend to do or think likewise. Thus, if it be possible for doctors to know this, a consequence would be an increase in the most productive – cost-effective – testings and interventions and a corresponding reduction in relatively wasteful ones. In this Information Age the implication is that the availability of user-friendly gnostic *expert systems* would enhance the efficiency of healthcare by inherently contributing to both quality assurance and cost containment in it (cf. propos. I – 1.5).

O. S. Miettinen, *Up from CLINICAL EPIDEMIOLOGY & EBM*,
DOI 10.1007/978-90-481-9501-5_8, © Springer Science+Business Media B.V. 2011

Proposition II – 3.3: To the visions about enhanced efficiency of healthcare that are based on the fundamental, obvious idea that the state of the knowledge-base of healthcare and access to it are critically important determinants of its level of efficiency (propos. II – 3.1, 2 above), an alternative has to do with provision for suitable *competition*: "The way to transform health care is to realign competition with *value for patients*. Value in health care is the health outcome per dollar of cost expended. If all system participants have to compete on value, value will improve dramatically" (ref. 1). However, in this course on 'clinical epidemiology' and EBM, the focus properly was on *knowledge* as a determinant of the efficiency of healthcare (as in propos. II – 3.1, 2 above), in contrast to addressing alignment of competition in this role.

References:
1. Porter ME, Teisberg EO. Redefining Health Care. Creating Value-Based Competition on Results. Boston: Harvard Business School Press, 2006; p. 4.
2. Porter ME. A strategy for health care reform – toward a value-based system. NEJM 2009; 361: 109–12.

The Dream of Universal Excellence in Healthcare

Proposition II – 3.4: When people choose a commercial flight from city A to city B, they choose the flight that is most convenient (as to schedule) or, perhaps, the most economical. In the choice of a flight people do not concern themselves with the pilot's and co-pilot's particular levels of expertise in providing a safe and otherwise trouble-free passage from A to B. They take it as a given that full competence/expertise – and overall excellence of practice – is universal among the pilots that airline companies hire; that airline pilots can always be expected to deliver the service that is the best that anyone in the profession can deliver, given the circumstances. The excellence of airline pilots is, in part, a matter of *professionalism* in the common meaning of practitioners in learned professions/occupations adhering to the dictates of selflessness, skill, and trustworthiness (ref.); but to this aviators have added *discipline* "in following prudent procedure and in functioning with others" (ref.). In medicine, however, "we hold up 'autonomy' as a professional lodestar, a principle that stands in direct opposition to discipline, [but it] hardly seems the idea we should aim for. It has the ring more of protectionism than of excellence" (ref.). Being able to presume universal excellence of practice among the doctors that agencies of healthcare hire would obviously be highly desirable, and not only from the consumers' vantage but from that of third-party payers as well. Thus, while it obviously is not the existing reality, universal excellence of healthcare should be brought about, if at all possible. The proximal challenge in bringing it about is to understand what would be involved.

Reference: Gawande A. The Checklist Manifesto. How to Get Things Right. New York: Metropolitan Books, Henry Holt and Company, 2009; pp. 182–3.

Proposition II – 3.5: Airline pilots naturally have a universally existing *motive* to represent excellence in providing for the safety of their clients, the passengers. For if they fail in this, they themselves may perish as well. In clinical medicine, by contrast, the doctor's health or survival is not jeopardized by failure to prevent the patient's demise. Unsurprisingly, thus, airline pilots have detailed definitions – carefully developed algorithms – of professional practice, and they follow these *norms* of practice willingly and closely, while doctors generally do not have detailed norms of clinical practice nor much desire to have these. But in clinical medicine, too, there should be detailed definitions of excellent practice – counterparts of the algorithms that airline pilots follow – and there should be, if needed, built-in artificial *incentives* for doctors to conform to those standards in their services (L. *servus*, 'slave') to their clients.

Proposition II – 3.6: The explicit, detailed definitions of excellent practices on the part of airline pilots are principally about their actions in the interest of their clients (while also about decision-relevant communication between the pilot and the co-pilot; refs. 1, 2). But in clinical medicine, the modern conception of professionalism no longer allows a doctor to act on behalf of a client on the presumption that the doctor on his/her own knows what action is in the client's best interest; in those decisions (s)he now needs to respect and defer to *patient autonomy* (ref. 3). Now, therefore, the definitions of excellence in clinical medicine – the stipulations of normative practices – are to be only about that which is in the essence of clinical medicine – the pursuit of facts bearing on gnosis (with the client's consent for this pursuit), the facts-conditional gnosis, and the teaching of the client accordingly (propos. II – 1.7) – but not about decisions.

References:
1. Gladwell M. Outliers. The Story of Success. New York: Little, Brown and Company, 2008; pp. 194 ff.
2. Gawande A. The Checklist Manifesto. How to Get Things Right. New York: Metropolitan Books, Henry Holt and Company, 2009; p. 125.
3. Participants in the Medical Professionalism Project. Medical professionalism in the new millennium. Lancet 2002; 359: 520–2. Also: Ann Int Med 2002; 136: 243–6.

Proposition II – 3.7: The dream of *universal excellence* in clinical medicine can be expressed this way: When a person consults a doctor (in the relevant discipline of clinical medicine), it does not matter who the doctor is: rather than a creative thinker subject to "cognitive errors" (ref. 1), the doctor inherently represents to the client access to – interface with – the knowledge that characterizes the top experts in the discipline (propos. I – 2.12, II – 2.11) and gives, in his/her teaching (propos. II – 1.7), the client the full benefit of this expertise (ref. 2).

References:
1. Groopman J. How Doctors Think. Boston: Houghton Mifflin Company, 2007.
2. Miettinen OS, Flegel KM. Elementary concepts of medicine: X. Being a good doctor: professionalism. J Eval Clin Pract 2003; 9: 341–3.

Proposition II – 3.8: "In a man's life dreams always precede deeds. Perhaps this is because, as Goethe said, 'Our presentiments are the faculties latent within us and signs of what we may be capable of doing … we crave for what we already secretly possess. Passionate anticipation thus changes that which is materially possible into dreamed reality' " – including, given that the requirements are in place, into universal excellence in clinical medicine (propos. II – 3.7 above).

> *Reference*: Marti-Ibañez F. The Epic of Medicine. New York: Clarkson N. Potter, Inc., 1961; p. xi.

Requirements for Universal Excellence in Healthcare

Proposition II – 3.9: For doctors generally to represent unsurpassed excellence in their respective disciplines of clinical medicine, three requirements need to be fulfilled: doctors in general are to have a motive for representing excellence; the practice-relevant knowledge (gnostic) of the top experts in each of the disciplines is to have been comprehensively and suitably codified; and the thus-codified knowledge is to be generally accessible ad hoc, as needed in the course of practice. (Besides, doctors generally are to have the requisite skills to well assemble the facts that constitute the ad-hoc inputs to the gnoses and to effectively teach the clients about the gnoses.)

Proposition II – 3.10: If the knowledge-base of a clinician's discipline is, to whatever extent, codified and readily accessible in the course of practice, the doctor tends to have more than a mere velleity to draw on it. For, the alternative would tend to be rather obvious malpractice, with a potential for adverse consequences to the doctor (as well as to the client). But to enhance the motivation to conform to the available knowledge, to the normative care (gnosis-related; propos. II – 3.6) implicit in this knowledge, third-party payers should – in the interest of quality assurance and cost containment – endeavor to cover normative care only (cf. propos. II – 3.5). (There must be no norms concerning decisions on behalf of clients who are adults, conscious, and sufficiently compos mentis to take the decisions; cf. propos. II – 3.6).

Proposition II – 3.11: To the extent that the knowledge-base of a clinician's discipline has been codified – in terms of gnostic probability functions (propos. II – 2.13) – it can be made readily accessible in the course of practice as a matter of applying already-existing information technology, by imbedding the knowledge-base in gnostic *expert systems*. And IT in the meaning of electronic health records will allow third-party payers to monitor, and on this basis enforce, conformity to the gnostic norms implicit in the expert systems (cf. propos. II – 3.10 above).

Proposition II – 3.12: For bringing about the dreamt-of universal excellence in clinical medicine (propos. II – 3.7), the requirement that at present is critically missing in whatever discipline of clinical medicine thus is only the more-or-less comprehensive *codification of experts' tacit knowledge in terms of GPFs*. But this requirement, too, can be met – and without major expense, even. It is thus time to get on with this.

Meeting the Missing Requirement for Universal Expertise in Healthcare

Proposition II – 3.13: The disciplines of clinical medicine are supposed to constitute a set of *learned* professions. Practice in each one of them thus is supposed to be, to the maximal possible extent, *knowledge*-based (propos. II – 1.14) – and not thinking-based, as Groopman describes it to be (ref. 1), or 'evidenced-based' (meaning: based on the practitioner's personal opinions about the implications of evidence; refs. 2, 3), as many now claim it should be (propos. I – 2.8).

> *References*:
> 1. Groopman J. How Doctors Think. Boston: Houghton Mifflin Company, 2007.
> 2. Evidence-Based Medicine Working Group. Evidence-Based medicine. A new approach to teaching the practice of medicine. JAMA 1992; 268: 2420–5.
> 3. Straus SE, Richardson WS, Glasziou P, Haynes RB. Evidence-Based Medicine. How to Practice and Teach EBM. Third edition. Edinburgh: Churchill Livingstone, 2005.

Proposition II – 3.14: As it is now, for no discipline of clinical medicine is the requisite knowledge-base – for setting gnostic probabilities – meaningfully and comprehensively codified, even though the pursuit and presumption of gnosis – at various levels of competence – goes on (in accord with the essence of medicine; propos. II – 1.7). The thus-far pre-eminent attempt at codification of expert knowledge failed (ref.). The reason for this was – according to the teachings in this course – that it wasn't understood what the necessary form of the knowledge for the purpose of its codification is (i.e., that of GPFs; propos. II – 2.13), to say nothing about not understanding how experts' tacit knowledge could be garnered for codification in that form.

> *Reference*: Wolfram DA. An appraisal of INTERNIST – I. Artif Intell Med 1995; 7: 93–116.

Proposition II – 3.15: Expert clinicians' gnosis-relevant general knowledge is not something they could make explicit in the form of GPFs or in some other general terms. Their knowledge is *tacit* in nature. They know about gnostic probabilities only ad hoc, in practice when gnostic challenges present themselves in their clinical encounters with clients; and in these instances, even, only in terms that are inconsistent across individual experts. Thus *the challenge* is to garner experts' tacit knowledge in the form of their typical ad-hoc beliefs about gnostic probabilities (propos. II – 2.11) and to give the pattern of these the form of GPFs – this on the premise that expertise on the topic actually exists.

Proposition II – 3.16: Given that expert clinicians know about gnostic probabilities in instances of gnostic challenge that actually occur in their practices, it follows that they equally know about them in hypothetical instances. From this it follows that insofar as experts' tacit knowledge about gnostic probabilities – for a particular object in a particular domain – exists, garnering it is most efficiently done on the basis of *hypothetical* instances presented to them; and the developmental challenge thus reduces to giving the thus-garnered tacit knowledge the form of a GPF addressing experts' typical beliefs (cf. propos. II – 3.15 above).

> *Reference*: Miettinen OS, Bachmann LM, Steurer J. Clinical diagnosis of pneumonia, typical of experts. J Eval Clin Pract 2008; 14: 343–50.

Proposition II – 3.17: For a domain of complaint-prompted pursuit of *diagnosis*, in the simple situation in which the basis of the diagnosis is only *two diagnostic indicators, both of them binary* (as in Assignments 4 and 6 in App. 3), the model for the requisite knowledge could be the 'saturated' one (Ass't 4) for each of the illnesses in the differential-diagnostic set (of possible underlying illnesses). For a given one of those illnesses, experts' tacit knowledge can be garnered in the form of that diagnostic probability function, DPF, validly and efficiently, as follows:

1. Each member of a panel of experts (dozens) is presented with a set of $N = 4$ hypothetical patients, one of each of the four possible kinds as for the diagnostic profile (representing a 'factorial design,' for efficiency); and for each of these, any given expert specifies what (s)he takes to be the most likely proportion of instances like this in general such that the illness in question is present (propos. II – 1.15).
2. The proportions/probabilities for the $N = 4$ 'patients' with their particular, different (X_1, X_2) profiles, where each X is an indicator $(0, 1)$ variate, are translated into the respective median (M) probabilities; and these, in turn, into their respective logits, $Y = \log[M/(1 - M)]$. The resulting dataset is constituted by the values of $\{Y, X_1, X_2\}_j$, $j = 1, 2, 3, 4$.
3. The saturated general linear model for the 'expected' value (i.e., the mean) of Y, involving $L = B_0 + B_1 X_1 + B_2 X_2 + B_3 X_3$, where $X_3 = X_1 X_2$, is fitted to the data. The result is $\hat{Y} = \hat{L}$, this linear compound involving the fitted/empirical values of the $\{B_i\}$ set $(i = 0, 1, 2, 3)$. This is the result for synthesizing with those on the same diagnostic function from other panels of experts. In the synthesis, the panel-specific results – the $\hat{B}$s in them – are weighted across the panels in proportion to the respective sizes of the panels, in averaging those panel-specific values for each of the parameters).

Proposition II – 3.18: For diagnostic situations more general than the extremely simple one addressed in proposition II – 3.17 above, novelties are prone to arise in respect to optimization (for efficiency) of the 'design matrix' in respect to both the univariate distributions and the joint distribution of the (set of) diagnostic indicators across the hypothetical patients addressed by the members of the expert panel. But otherwise the process of garnering the panel members' tacit knowledge remains in principle the same as in the proposition above. Obviously needed is a larger number of different profiles represented by the 'patients,' all different in respect to their profiles but by no means representing all of the possibilities.

Proposition II – 3.19: As for the specifics left unaddressed by proposition II – 3.18 above, examination of the still relatively simple diagnostic situation in Assignment 5 (App. 3) is instructive. Let us define

X_1 as indicator of moderate symptom (1 if present, 0 otherwise),
X_2 as indicator of severe symptom,
X_3 as (the numerical value of) the time of the test (with T_0 the time of the symptom's onset), and
X_4 as (the numerical value of) the level of the test result;

and let us take it that the model for the logit of the diagnostic probability for the illness at issue was designed to involve

$$L = B_0 + B_1X_1 + B_2X_2 + B_3X_3 + B_4X_4 + B_5X_3^2 + B_6X_4^2 + B_7X_3X_4 + B_8(X_3X_4)^2.$$

The design matrix again specifies the joint distribution of the diagnostic indicators, and thus of X_1 through X_4, in the hypothetical case profiles presented to the panel of experts. In designing the matrix, the concern again is to maximize the efficiency of the project. In the development of the design matrix, an obvious novelty now has to do with the quantitative scales of X_3 and X_4, the design of their univariate distributions and then the extension of the factorial design for the joint distribution of X_1 and X_2 (as in propos. II – 3.17) to that for X_1 through X_4.

Proposition II – 3.20: If the model in proposition II – 3.19 above did not involve the square terms for X_3 and X_4 (and for their product), the univariate distributions of these two variates would be efficient in terms of the 'two-point design' of equal allocation of the 'patients' to the extremes of their respective ranges (realistically, in the domain at issue). But given the allowance for curvature in these relations in the adopted model, needed is the 'three-point design,' that is, equal allocation to the three points constituted by those extremes together with a point in the middle of the range. Now the four variates jointly specify $2 \times 2 \times 3 \times 3 = 36$ different profiles, and the efficient design matrix specifies 36 'patients,' one of each type of the diagnostic profile.

Proposition II – 3.21: That viewpoint of efficiency maximization by means of the orientational principles of the two- and three-point designs and those of factorial designs in general is not to be viewed as a generally desirable one to adopt. An obvious exception has to do with (near-)pathognomonic elements in the manifestational segment of the profile, either positively (rule-in) or negatively (rule-out) pathognomonic. The significance of these indications can be established with very few instances of them in the set of 'patients,' and they really are to be made so uncommon that the vast majority of the instances represent the subdomain in which the diagnosis is challenging. Thus, perhaps only two of the 'vignettes' are to have a given positively pathognomonic indication in the profile, and the same applies to negatively pathognomonic profiles. Apart from these extremes, lopsided distributions may be desirable for various other diagnostic indicators as well. For example, if for the diagnosis of acute myocardial ischemia the chief complaint is acute 'dyspnea, chest pain, or both,' with various descriptors of the pain but not of the dyspnea, there is greater interest in the instances of chest pain with its various particulars, and these instances therefore should be more common than those of dyspnea in the set presented to the experts. (Cf. App. 5.)

Proposition II – 3.22: With the design matrix and, thereby, the set of 'patients' defined with a view to appropriate univariate distributions of the diagnostic indicators and maximal possible independence of distributions among them, the number of parameters in the model may exceed the number of 'patients' (for each of which the typical diagnostic probability of the expert panel's members is documented), and one consequence of this is that the model cannot be fitted to the data in the usual way. The solution to this, in its simplest form, is to partition the model by taking the linear compound to be $L = B_0 + \Sigma_k B_k L_k$, where each component L_k is based on a particular subset of the indicators, with these sets non-overlapping yet jointly all-inclusive as for the indicators and Xs at issue. Each $L_k = (B_0 + \Sigma_i B_i X_i)_k$ is fitted, separately, to the data, and the value of the fitted L_k is calculated for each 'patient,' for each k. And finally, that overall model, formulating the typical probability's logit as $L = B_0 + \Sigma_k B_k L_k$, is fitted to the data (realizations of L_k, $k = 1, 2, \ldots$) – and reduced to the form in which $L = B_0 + \Sigma_i B_i X_i$. Involved could be L_1 and L_2 based on risk and manifestational indicators, respectively.

Proposition II – 3.23: Another challenge in this context is that the familiar factorial design in the context of increasing number of diagnostic indicators soon becomes impracticable on account of the large number of 'patient' profiles that need to be specified (to retain the orthogonality). In this, very helpful has been the group, or field, theory – that which Évariste Galois feverishly wrote down the night before his young life was to end in a hopeless duel in 1832 – which has profoundly advanced both mathematics and physics (ref. 1). It has also provided the basis for a stunningly powerful – and elegant – *extension of the factorial design* in optimizing the design matrix for efficiency in industrial experimentation, for which the extension was developed (ref. 2). With this extension of the factorial design as a critical input, the principles of the matrix design are elaborated further, in the context of an example, in Appendix 5.

References:
1. Berlinski D. Infinite Ascent. A Short History of Mathematics. London: Weidenfeld & Nicolson, 2005; pp. 85 ff.
2. Plackett RL, Burman JP. The design of optimum multifactorial experiments. Biometrika 1946; 33: 305–25.

Proposition II – 3.24: If the result is meant to be applied without its synthesis with other results, it needs to be adjusted for 'overparametrization' (too many parameters in proportion to the number of datapoints), for the 'overfitting' associated with this. One way to do the adjustment is the '*leave-out-one*' method:

1. Of the N different 'patient' presentations to the members of the panel – more than the number of parameters in the model – one is left out in deriving the empirical L, the result being $\hat{L}'$ (while it was $\hat{L}$ on the basis of all N of the presentations). The value $\hat{Y}'$ of the $\hat{L}'$ is calculated for the 'patient' that was left out, and it is paired with the corresponding value, Y, for the logit of the experts' actual diagnostic-probability median for the left-out 'patient.' This process is carried out for each of the N – 1 other presentations as well. The result is N realizations for the $(\hat{Y}', Y)$ pair.

2. The model expressing the mean of Y as $B_0' + B_1'\hat{Y}'$ is fitted to the N datapoints on $(\hat{Y}', Y)$. The result, $\hat{B}_0' + \hat{B}_1'\hat{Y}'$, generally involves $\hat{B}_1' < 1$ as a manifestation of the bias in $\hat{L}$ – propensity of the high values to be too high, the low values too low, and thus to exaggerate the discriminating information in the Xs. This calls for adjustment of $\hat{L}$ in terms of the 'regression toward the mean,' represented by the mean of Y as a function of $\hat{Y}'$, the adjustment being substitution of $\hat{L}^* = \hat{B}_0' + \hat{B}_1'\hat{L}$ for $\hat{L}$. An alternative to this adjustment of $\hat{L}$ is to derive $\hat{L}^*$ by averaging the values of $\hat{B}_0'$ and $\hat{B}_1'$ across the N values for each of these. The bias-adjusted diagnostic probability function (empirical) thus becomes $P = 1/\left[1 + \exp(-\hat{L}^*)\right]$ (cf. propos. II – 2.14).

Proposition II – 3.25: Pertaining to *etiognosis*, experts' tacit knowledge cannot be about etiognostic probability as such; it must be about etiognosis-relevant causal rate-ratios (propos. II – 1.29). In the design matrix the Xs again represent the explanandum's – here the rate ratio's – determinants per se (exclusive of, e.g., product terms). The principles of optimization of the design matrix (for efficiency) are the same as in the context of garnering experts' diagnostic knowledge in the appropriate – functional – form. A linear model is designed for the logarithms of the experts' typical – median – case-specific best surmises about the magnitudes of the etiognosis-relevant causal rate-ratios, this model is fitted to the data, and the fitted function is exponentiated. Overfitting tends to be less of an issue than in the diagnostic context, as fewer parameters tend to be involved in the rate-ratio function.

Proposition II – 3.26: As for *prognosis*, the point of principal note is that experts' tacit knowledge – different from what may be directly addressed in prognostic research – never is about incidence density; it always is about prognostic probabilities themselves, again specific to particular instances (now as to prognostic profile, intervention and a particular period of, or point in, prognostic time). Thus, even for the probabilities of an event the model is to be logistic, for probability, rather than log-linear for incidence density. Overfitting generally is, in its significance, intermediate between those in the diagnostic and etiognostic contexts.

PART III
THEORY OF CLINICAL RESEARCH

III – 1. ENHANCEMENT OF PRACTICE BY CLINICAL RESEARCH
Evidence as the Product of Clinical Research
Evidence as a Supplement to a Clinician's Experience
Evidence in the Advancement of Clinical Knowledge
Evidence in the Enhancement of Clinicians' Efficiency
Priority-Setting for Quintessentially 'Applied' Clinical Research

III – 2. INTRODUCTION INTO DIAGNOSTIC CLINICAL RESEARCH
The Nature of the Results of Diagnostic Clinical Studies
The Genesis of the Results of Diagnostic Clinical Studies
The Quality of the Results of Diagnostic Clinical Studies
Screening Studies as Exceptions in Diagnostic Clinical Research

III – 3. INTRODUCTION INTO ETIOGNOSTIC CLINICAL RESEARCH
The Nature of the Results of Etiognostic Clinical Studies
The Genesis of the Results of Etiognostic Clinical Studies
The Quality of the Results of Etiognostic Clinical Studies
The 'Cohort' and 'Trohoc' Fallacies in Epidemiologists' Etiologic Studies

III – 4. INTRODUCTION INTO PROGNOSTIC CLINICAL RESEARCH
The Nature of the Results of Prognostic Clinical Studies
The Genesis of the Results of Prognostic Clinical Studies
The Quality of the Results of Prognostic Clinical Studies
On Guidelines for Reporting on Clinical Trials

III – 1. ENHANCEMENT OF PRACTICE BY CLINICAL RESEARCH

Evidence as the Product of Clinical Research

Proposition III – 1.1: Given an object of inquiry of the scientific – abstract-general – sort for the knowledge-base of clinical medicine, a single piece of research on it – a clinical *study* in this meaning – is not reasonably construed as a project to produce an/the answer to a question about it, much less a/the conclusion about it. For, as the term 'research' – re-search – suggests, any object of inquiry in empirical science commonly needs to be studied with successive *replications*, for the development of more-or-less firm knowledge about it. Realistically, therefore, any given study in that succession of studies is to be understood to be a project to make a contribution to the aggregate of evidence concerning the object of inquiry – the abstract-general truth (invariant by place and time) that is at issue, such as the (substantive) content of a gnostic probability function, GPF, of a predesigned form.

Proposition III – 1.2: A study in the succession of studies on a given object of inquiry – not only the initial one but each of its replications just the same – is a piece of so-called *original* research. In research on GPFs there also is a need for *derivative* research in the meaning of identifying all of the already produced evidence on the object of study (GPF) at issue and producing a synthesis of this evidence. A piece of this latter type of research is now commonly referred to as a 'systematic review' (of the original research), and the statistical synthesis of the numerical evidence (from the original studies judged to be valid) as 'meta-analysis.'

Proposition III – 1.3: The larger is the number of replications of the initial study on the object of inquiry at issue, the greater is the importance of the derivative research relative to any given one of the original studies on the object, but it always is – in principle at least – more important than any one of the original studies, the evidence from which it synthesizes.

Proposition III – 1.4: Like the evidence from original studies, that from derivative studies also is incompletely reproducible in replications of these studies; and the less reproducible the result from a study is, the more important is replication of the study – in derivative research even without any interim increase in the evidence produced by original studies.

Proposition III – 1.5: The *evidence* from a piece of original research on a GPF, and from a piece of corresponding derivative research just the same, is not constituted solely by its published *result* – the empirical values of the parameters in the object of study. The result is rather meaningless if divorced from the rest of the evidence, namely, publicly documented *genesis* of the result. This latter segment of the evidence from a study has to do with the study's methods design and the deviations from this in the execution of the study, supplemented by whatever insights the investigators have into the study process or data that might have further bearing on the result's interpretation. The evidence is *objective* in these terms – agreeable by all concerned, as to what it is.

Proposition III – 1.6: The result's genesis determines its quality in reference to the object of study. This quality is constituted by the result's degree of *validity*: freedom from tendency to be in error (to deviate from the truth about the object of study, in the domain of it) in a particular (commonly knowable) direction to an unknowable extent – freedom from bias, that is. The result's degree of *precision* or reproducibility – a matter of the quantity, rather than quality, of information represented by the result – is determined by its genesis in the study's size together with its efficiency. Different from the result's bias, its (im)precision can be assessed statistically (in terms of 'standard error' or a 'confidence interval').

Proposition III – 1.7: Evidence with even a high degree of both validity and precision can be seriously *misleading* on account of unrecognized flaws in the study's object design. One eminent example of this, among many others, presumably is the research (by randomized trials) purported to have shown ineffectiveness of antioxidant supplementation of diet in the prevention of cancers. The problem with this evidence is the failures, in the studies' object designs – insofar as the investigators even think about this – to appreciate the presumably very long time lag – decades – from the (initiation of) the hypothesized pharmacological retardation of the pathogenetic process of cumulative genetic damage (from free radicals and 'toxic oxygen') to the appearance of overt cancer, intended to be prevented or at least delayed by the 'chemoprevention.' Another eminent example is epidemiologists' research on screening for a cancer, again with commonly negative results on account of failure to suitably address issues in the object designs, and methods designs subordinate to these, in such studies as epidemiologists have been and still are committed to in this research – instead of leaving the knowledge-base of screening – pursuit of early, preclinical diagnosis – to clinicians to develop (cf. sect. IV – 2).

Proposition III – 1.8: '*Strength*' of the evidence, notably of the aggregate of the evidence from derivative research, is an unworthy concept cultivated by 'clinical epidemiologists.' It is expressed in terms of ordinal scales that have no proper foundation in the result's validity and precision in reference to the object of study such as it actually was, nor in the result's relevance per its form (that of the object of study; cf. propos. III – 1.7 above).

Proposition III – 1.9: Mastery of the theory of clinical research (of the quintessentially 'applied' variety; propos. I – 2.5) is distinctly more important in derivative research than in original research. For, the selection of original studies into the synthesis of their results involves judgments about their quality (admissibility/relevance of the object of study, validity of the methods) that may override those of the original investigators and, notably, those of their 'peer reviewers' as well. Unwittingly perhaps, but nevertheless regrettably, *standards* for original research are now being set by (the criteria for original studies' inclusion in) derivative research – commonly conducted by self-appointed groups with no special expertise in the theory of quintessentially 'applied' clinical research. Expertise in 'systematic reviews' now is commonly involved, but this is an essentially vacuous succedaneum for that relevant expertise on clinical research, original first and foremost and derivative only secondary to this.

Evidence as a Supplement to a Clinician's Experience

Proposition III – 1.10: Remarkable though it is, in this Information Age in particular, the knowledge-base of whatever discipline of clinical medicine (for gnoses in it) is, still, nowhere comprehensively and truly meaningfully codified (à la propos. II – 2.13); and hence, much reliance is placed, still, on the role of a doctor's *personal experience* as a source of (a semblance of) the requisite knowledge.

Proposition III – 1.11: Personal experience is not evidence in the meaning of this in science: even if of the form of scientific evidence, it lacks the objectivity of evidence from research – the quality of the latter that there can be general agreement by all concerned about what the evidence is (propos. III – 1.5). Nor does the personal experience of a doctor produce for him/her actual medical knowledge in the intersubjective meaning of experts' typical beliefs about the magnitudes of gnostic probabilities (propos. II – 2.11). It can produce subjective beliefs – personal opinions – only.

Proposition III – 1.12: When a doctor's recollection of and inference from personal experience with a given type of gnostic challenge is supplemented by his/her familiarity with – and personal evaluation of and inference from – *evidence* from clinical research, his/her beliefs about those gnostic probabilities are prone to change; but despite the objective input into them, the updated beliefs still are eminently subjective. They still do not represent knowledge in the (relatively relaxed) meaning of experts' typical beliefs (propos. II – 2.11).

Proposition III – 1.13: *Evidence-based medicine* – whether simply in the meaning of a doctor basing (gnosis in) his/her practice on his/her personal evaluation of and inference from evidence as a supplement to his/her personal experience to whichever extent, or in the meaning of this evaluation and inference supplemented by the rest of the body of EBM doctrines (ref.) – is medicine based on subjective beliefs (propos. III – 1.12 above) and thereby not even professional medicine (propos. I – 5.14), much less scientific medicine (in which the theoretical framework is rational and scientific knowledge is deployed – for gnosis – in such a framework; propos. I – 2.9).

> *Reference*: Straus SE, Richardson WS, Glasziou P, Haynes RB. Evidence-Based Medicine. How to Practice and Teach EBM. Third edition. Edinburgh: Churchill Livingstone, 2005.

Evidence in the Advancement of Clinical Knowledge

Proposition III – 1.14: That evidence from quintessentially 'applied' medical research (propos. I – 2.5) is prone to change experts' typical beliefs about gnostic probabilities is not only intended by the research but also a given (propos. III – 1.12 above); but this does not mean that such evidence inherently advances anyone's gnosis-relevant knowledge (in the meaning of experts' typical beliefs; propos. II – 2.11). For, such evidence does not, in itself, imply what the evidence-informed beliefs of experts (regarding gnostic probabilities in particular instances from the domain of the evidence) typically are; evidence-informed experts themselves, even, do not inherently have this knowledge. Accumulation of evidence thus is not tantamount to advancement – or even existence – of evidence-based/evidence-enhanced knowledge.

Proposition III – 1.15: Evidence advances practice-relevant clinical knowledge (gnostic) insofar as there is codification/documentation of experts' typical evidence-advanced beliefs in the form of GPFs (propos. II – 3.15, etc.), and as a practical matter insofar as the ad-hoc implications of these GPFs are made accessible to doctors via expert systems (propos. II – 3.11).

Proposition III – 1.16: Evidence has its optimal impact in the advancement of the knowledge-base of practice if the expert panels involved in the codification of the knowledge (propos. II – 3.16, etc.) are top clinicians (gnosticians) on the topic at issue and also familiar with all of the available evidence – original and derivative – on the object of gnosis at issue in the domain at issue (e.g., presence/absence of illness I as the object of diagnosis in the presentation domain D), and two additional conditions of expertise also obtain: The panel members – clinical academics – know and understand the theory of the relevant genre of gnostic research (diagnostic, say) and are, thereby, qualified to critically assess the appropriateness/relevance of the object of study and the quality of the empirical GPF(s) (propos. I – 2.13); and this assessment is supplemented by opportunity to evaluate the solely evidence-based empirical probability values (diagnostic or prognostic) in the context of particular 'patients' (hypothetical) from the domain at issue – by these values being on display as additions to the usual 'facts' on the hypothetical patients presented to the panel members for probability-setting (à la propos. II – 3.16, etc.).

Proposition III – 1.17: In codifying (a given segment of) the knowledge-base of clinical medicine (à la propos. II – 3.16, etc.), the *role of evidence should be optimized* in accordance with proposition III – 1.16 above, in respect to both the panels of experts and the 'patient' and evidence presentations to its members, and also in terms of the translation of the expert inputs into GPFs. Implementation of this precept remains challenging, however, on account of paucity of both genuine experts (propos. I – 2.15) and the appropriate type of evidence (sect. IV – 2).

Evidence in the Enhancement of Clinicians' Efficiency

Proposition III – 1.18: Once experts' evidence-enhanced tacit knowledge (in ad-hoc applications) has been garnered in the form of GPFs (characterizing experts' typical beliefs) and also has been made accessible to doctors as needed in the course of practice (by means of expert systems), evidence presumably elevates the general level of efficiency of clinicians' practices by improving the *quality* of clinical care and thereby also serving to contain its *cost* (propos. II – 3.2).

Proposition III – 1.19: For evidence to have its maximal impact in quality assurance and cost containment of clinical care, optimization of its impact in the advancement of knowledge (propos. III – 1.16–17 above) and making this knowledge accessible in the course of practice (propos. III – 1.18 above) needs to be supplemented by measures to enhance the deployment of this readily-available evidence-enhanced knowledge (propos. II – 3.5). In maximizing conformity of practice with the implicit norms embedded in the expert systems, professional societies have an educational and disciplinary responsibility and role, while a quasi-regulatory role can be played by third-party payers of the care (propos. II – 3.10–11).

Priority-Setting for Quintessentially 'Applied' Clinical Research

Proposition III – 1.20: Whereas the overall mission in quintessentially 'applied' clinical research should be understood to be the production of evidence for the advancement of the knowledge-base of clinical practice; whereas there is very much to do in this vein (especially now that almost nothing has been meaningfully done); and whereas the research generally is quite demanding of both time and resources, priority-setting among its possible topics is important in an effort to maximize the *cost-effectiveness* of the research. (The burdensome character of the research is in sharp contrast to the nature of the developmental work of garnering experts' tacit knowledge in the form of GPFs – for expert systems.)

Proposition III – 1.21: From the vantage of any given clinical investigator, rational priority-setting among possible topics for quintessentially 'applied' clinical research naturally can be internal to his/her particular discipline of clinical medicine; and it may well also be internal to his/her 'specialty' among the diagnostic, etiognostic, and prognostic genera of such research.

Proposition III – 1.22: In diagnostic clinical research – all of which is to address situations of decision-making about action (testing or intervention) – priority belongs to decision situations that are relatively common, in the context of which the choice of action is prone to be particularly urgent and consequential, and concerning which the existing degrees of expertise among the top experts remain particularly wanting. Those situations actually are quite common, and so also are serious consequences of misdiagnoses in them (ref.).

> *Reference*: Newman-Toker DE, Pronovost PJ. Diagnostic errors – the next frontier for patient safety. JAMA 2009; 301: 1060–2.

Proposition III – 1.23: In etiognostic clinical research, the highest priority generally belongs to study of etiogenesis of serious adverse events (or states) in respect to such uses of medications as are commonplace in one's particular discipline of clinical medicine. For, etiognosis in respect to such etiogenesis is critically important for prevention of future recurrences of them in the same patients, more-or-less idiosyncratic reactions in particular (by means of withdrawal of the medication's current use when found to be etiogenetic and never using it in the same patient again). In some disciplines of clinical medicine, study of the etiogenesis of some of the illnesses of inherent concern in it may deserve priority on the basis of concern for clinical prevention of their occurrence – in analogy with epidemiologists' etiologic/etiogenetic research with a view to community-level preventive medicine.

Proposition III – 1.24: Prognostic clinical research generally is focused, quite justifiably, on intervention-prognosis. In this the need for evidence generally is greatest for prognosis about the prophylactic effectiveness of chronic interventions, for not only would the evidence-advanced knowledge about effectiveness be used in important decisions (by doctors' clients), but tacit knowledge of this type tends not to accrue on the basis of mere personal experience with the interventions.

III – 2. INTRODUCTION INTO DIAGNOSTIC CLINICAL RESEARCH

The Nature of the Results of Diagnostic Clinical Studies

Proposition III – 2.1: While objectively empirical in content, the *form* of a result of a diagnostic study naturally is to be that of an element in the requisite knowledge-base of diagnosis proper – diagnostic probability-setting in the face of a given diagnostic profile – or of the decision about invocation of (further) diagnostic testing.

Proposition III – 2.2: For (the advancement of) knowledge relevant to diagnostic probability-setting, the form of the study result is to be that of a diagnostic probability function, DPF, for a defined domain of client presentation (propos. II – 2.14–16). An empirical DPF pertains to the knowledge-base of pre-test or post-test diagnosis according as the result of the test at issue isn't or is among the diagnostic indicators accounted for in the function (propos. II – 2.17), given that the indicators shared by these two functions define a 'decision node' in respect to the test's use.

Proposition III – 2.3: An empirical post-test DPF pertains to the knowledge-base of decision-making about the test's use, a preliminary aspect of this. At issue is determination of the range of possible post-test probabilities conditional on the pre-test profile; and in particular, determination of the range of the test's results which, if any, would provide for 'conclusive' diagnosis in the meaning of a practical rule-in or rule-out diagnosis, as its probability range is set by the diagnostician (propos. II – 2.17).

Proposition III – 2.4: For the knowledge-base of decisions about a test's use, additional results of a diagnostic study may be functions that address the probabilities of various potentially 'conclusive' ranges of the test's result, conditionally on the pre-test profile. For, in conjunction with the post-test DPF, generally needed are test-result probability functions, TRPFs, for various ranges of the test result, functions that express how their probabilities depend on the pre-test profile (propos. II – 2.18). A particular one of these is relevant in a given decision (depending on what the test-result range for 'conclusive' diagnosis is in the instance at issue; cf. propos. III – 2.3 above).

O. S. Miettinen, *Up from CLINICAL EPIDEMIOLOGY & EBM*,
DOI 10.1007/978-90-481-9501-5_10, © Springer Science+Business Media B.V. 2011

The Genesis of the Results of Diagnostic Clinical Studies

Proposition III – 2.5: The general process of diagnostic clinical studies of the *original* sort, constituting the genesis of the study results (of the type outlined above) and thereby the substance of the evidence from the studies jointly with the results (propos. III – 1.5), should be understood to involve these sequential elements in the genesis of the *study series*, suitably documented, on an instance-by-instance basis:

1. Identification of an instance of the domain of the diagnosis at issue and, hence, of the domain of the function(s) that is (are) the object(s) of the study.
2. Decision about solicitation, in the identified instance, of consent to participate in the study (upon the person having been thoroughly informed about what the participation would entail, most notably as to experimental testing (in respect to the diagnostic indicators in the object of study), if any is involved, and in any case as to how the truth about the presence/absence of the illness at issue would be determined, should the acquired facts call for this (see below).
3. Given (the solicitation and attainment of) informed consent, documentation of the realizations of the diagnostic indicators involved in the object(s) of study, this in accordance with the study protocol's definitions of the empirical – operational – scales of the indicators (reflecting concern for objectivity and truth more than may be routine in practice).
4. Given (informed consent and) the documented diagnostic profile(s), decision about whether to determine the truth regarding the presence/absence of the illness at issue.
5. Given the decision to do this, determination and documentation of the truth about the presence/absence of the illness at issue – and, thereby, inclusion of the instance in the study series (of select and suitably documented instances from the study domain).

Proposition III – 2.6: Given the study series of suitably documented instances from the study domain, the data on these are translated into realizations of the primary statistical *variates* involved in the object of study ($Y, X_1, X_2, \ldots ; Y = 1$ if the illness was present, 0 otherwise), pre-designed all the way to the (form of the) logistic DPF (propos. II – 2.14), with these variate data possibly supplemented by realizations for variates addressing the distribution of a test's result ($Y_i = 1$ if the result is in range R_i, 0 otherwise; cf. propos. II – 2.18, III – 2.4).

Proposition III – 2.7: The designed object *functions* are fitted to the data, without (data-driven, stepwise or other) reduction, including with adjustment for over-parametrization/overfitting if at issue is the first study on the object and its result might be applied as such, before synthesis with those from other studies (propos. II – 3.24). The results without the adjustment are reported regardless – for the purposes of derivative studies on the objects of study. (For the parameters of the object functions, reported are the fitted values together with their standard errors.)

Proposition III – 2.8: The genesis of the result of a *derivative* clinical study for diagnosis (on a given, pre-designed object of study) involves these elements:

1. Identification of all original studies on the function that is the object of the study.
2. Selection, from among those original studies, of the ones to be accounted for in the derivative study.
3. Synthesis of the results from the selected original studies as a matter of calculating the information-weighted averages of the study-specific empirical values for each of the parameters in the object function, those information-proportional weights being the inverses of the squares of the respective standard errors.

The Quality of the Results of Diagnostic Clinical Studies

Proposition III – 2.9: The quality of a given result from a diagnostic clinical study has, most broadly, two principal determinants: the study's *objects design* and *methods design*, with the degree of success in the designed methods' execution having an additional role (propos. III – 1.5).

Proposition III – 2.10: There continues to be much confusion about the quality of the results of diagnostic clinical research in respect to the consequences of the studies' objects designs, determining the results' generic nature. Most of the confusion has been adduced by radiologists, while some of it has been both adduced and propagated by 'clinical epidemiologists.' But there should be no confusion about this: rationality dictates that the results be (empirical) diagnostic probability functions, DPFs, possibly supplemented by test-result probability functions, TRPFs (propos. III – 2.2–4).

Proposition III – 2.11: The *domain* of the object function(s) is to be defined in terms that have universal meaning in respect to the substance of medicine. It thus must not be one of suspicion of the illness being present (referring to the minds of diagnosticians), nor can it rationally be one of patients referred for the diagnosis (with actions of diagnosticians domain-defining). It must be defined on the basis of relevant facts about the presentation per se; and just as these do not include anything about the doctor's cognitions or actions, they also do not include anything about such incidental matters as the type of practice (as to its 'setting' or 'specialty,' say), which again in no relevant (and universally meaningful) way describes the types of particular client presentations (to whatever practices).

Proposition III – 2.12: Whatever is the study object's admissible domain, the subdomains-defining set of diagnostic *indicators* must fully account for the diagnosis-relevant reasons why instances from the domain come to diagnostic attention. As for the role of diagnostic testing in this, the domain may be designed to be one in which no testing preceded the presentation; but otherwise the background testing and its result(s) are to be accounted for in the definition of subdomains (by means of the diagnostic indicators in the objects of study).

Proposition III – 2.13: The empirical scales of the diagnostic indicators (including in their role in the definition of the functions' domain) – and hence the terms in which the indicators' realizations enter the diagnostic profiles (and domain recognitions) – are to represent *genuine facts*. This is of particular note in respect to the indicators that generally are documented anamnestically (from interview of the study subject); but even those that have to do with findings from physical examination can be subject to error. (A function based on genuine facts can be meaningfully used in the face of potential factoids, as a matter of exploring the implications of various possible [sets of] facts, while an empirical function based on mere factoids is of little or no use.)

Proposition III – 2.14: *The set* of diagnostic indicators in a result for a particular decision node (re action) is to be comprehensive, encompassing all of the indicators that reasonably could be accounted for in that situation and also could be considered having (marginal) relevance in it. For, anyone who postulates such relevance is left unsatisfied with a result that does not address the potential supplementary/marginal informativeness of some left-out indicator(s).

Proposition III – 2.15: While the objects design of a diagnostic study thus determines the scientific admissibility and, thereby, the 'applied' relevance of each of the objects of study, the quality of the result on an admissible-and-relevant object of diagnostic study is a matter of the methods' degree of *validity* – freedom from propensity to introduce bias into the probability estimates that use of the function (at face value) produces.

Proposition III – 2.16: Critical in the validity-assurance for a diagnostic study of the original type are the inclusions of instances (from the study objects' domain) in the study series – specifically, freedom from *selection bias* in these inclusions. The need is to assure, by suitable selection, two qualities for what ultimately is the study series of instances: that it does not include ineligible instances, ones in which the 'facts' are (to some extent) suspected to be incorrect, or instances in which the determination of the fact about the presence/absence of the illness got to be influenced by correlates of this other than the diagnostic indicators accounted for in the object(s) of study; and that it does include all of the eligible instances in which the truth about the presence/absence of the illness was determined. Thus, the fact-finding really should not proceed from the profile documentation to determination of the presence/absence of the illness if there is incomplete assurance of the correctness of the profile-constituting 'facts'; and the decision to ascertain the truth about the presence/absence of the illness, if taken, really should be followed by unconditional success in this (and, thereby, inclusion in the study series of instances without a role for latent correlates of the truth at issue).

Proposition III – 2.17: If the study is a derivative one, its result's validity requires inclusion of valid original studies only, and independently of their result(s). The inclusions' independence of the study result(s) generally requires inclusion of all of the valid studies, unpublished ones included. Maximization of precision also requires this inclusiveness.

Proposition III – 2.18: The precision of the object function's parameter values in the study result, or the precision of the probability estimates produced by the function, is not a matter of the result's quality. Rather, it reflects the quantity of information embodied in the study result, consequent to the study's efficiency together with its size (cf. propos. III – 1.6).

Proposition III – 2.19: A major determinant of the *efficiency* of an original diagnostic study is the way in which decisions about the determination of the truth about the presence/absence of the illness are taken, notably the extent to which selectivity in these determinations shapes the (joint) distribution of the diagnostic indicators toward a low degree of collinearities in this (which is such an eminent pursuit also when hypothetical instances are used in garnering experts' tacit knowledge in the form of DPFs; propos. II – 3.23).

Screening Studies as Exceptions in Diagnostic Clinical Research

Proposition III – 2.20: A diagnostic study need not serve diagnosis in the context of a complaint. The domain for a diagnostic study can be one of no complaint, as an *apparently healthy* person may need rule-out diagnosis of a particular illness (for, say, occupational or insurance purposes); or (s)he may seek *rule-in* diagnosis about a particular illness (cancer, notably) – that is, pursuit of this by means of *screening*. In the latter case the aim, more specifically, is to achieve early, latent-stage diagnosis (rule-in) about the illness, thereby providing for early, more effective (and otherwise more attractive) treatment.

Proposition III – 2.21: Given a *regimen* of screening for a particular illness (for the pursuit of latent-stage rule-in diagnosis about it), the principal object of the requisite knowledge and, hence, of screening research is the resulting *diagnostic distribution* – of diagnosed cases, according to major prognostic indicators (stage of cancer, say).

Proposition III – 2.22: While screening for an illness – a cancer, notably – generally is a multidisciplinary topic of *clinical* diagnosis (propos. II – 1.24), *epidemiologists* are in the habit of thinking about it as a matter of the initial testing in that pursuit, taking this testing to constitute community-level preventive intervention (to reduce mortality from the illness) and as a matter to be governed by public policy (different from what is normal in respect to clinical medicine, in contrast to community medicine). The consequences for screening research – and practices – have been, and are, very sad.

References:
1. Miettinen OS, Henschke CI, Pasmantier MW, et alii. Mammographic screening: no reliable supporting evidence? Lancet 2002; 359: 404–5.
2. Miettinen OS. Curability of lung cancer. Expert Reviews of Anticancer Therapy 2007; 7: 399–401.
3. Miettinen OS. Screening for a cancer: a sad chapter in today's epidemiology. Eur J Epidemiol 2008; 23: 647–53.
4. Miettinen OS. Screening for a cancer: thinking before rethinking. Eur J Epidemiol 2010; 25: 365–74.

III – 3. INTRODUCTION INTO ETIOGNOSTIC CLINICAL RESEARCH

The Nature of the Results of Etiognostic Clinical Studies

Proposition III – 3.1: Whereas the knowledge-base of clinical *dia*gnosis is entirely acausal as for the determinants of the diagnostic probabilities (even though the illness at issue is a potential cause of the manifestational profile, and causes of the illness have a role in the risk profile), the knowledge-base of *etio*gnosis is expressly – and explicitly, per the 'etio' prefix of the term (adduced only as recently as in 1998; ref.) – about causality.

> *Reference*: Miettinen OS. Evidence in medicine: invited commentary. CMAJ 1998; 158: 215–21.

Proposition III – 3.2: Aristotle distinguished among four types of 'cause' (as *aitia* – on which the 'etio' prefix is based – questionably has been translated) – material, formal, efficient, and final. An antecedent constituting or completing what we think of as a/the sufficient cause is what he meant by 'efficient cause' – of a phenomenon that has occurred or does occur – a case of an illness, say. Causation in this post-hoc, retrospective, *explanatory* sense constitutes a topic very different from causation in the prospective, course-altering, anticipatory sense – of intervention-prognosis, say.

Proposition III – 3.3: The essential result of an etiognostic clinical study addresses a *causal rate-ratio* (RR) – rate of the occurrence of the illness/sickness given a risk factor's index category (representing a potential cause as the antecedent) divided by a comparable counterpart of this with the reference category (representing the alternative to the potential cause in the causal contrast). For, an empirical value for the *etiogenetic fraction*, EF = (RR – 1) / RR, and thus for the etiognostic probability, is implied by this (propos. II – 1.16, 29).

Proposition III – 3.4: If the antecedent can be preventive in some instances while causal in others, with the latter instances more common (so that causal RR > 1.0), then that RR-based measure of EF is but the lower bound for the etiognostic probability at issue.

Proposition III – 3.5: The etiognostically relevant RR result of an etiognostic clinical study is to be subject to quantitative causal interpretation (in reference to a defined domain), and it commonly also needs to represent the empirical RR as a function of the temporal (and perhaps other) particulars of the generic antecedent at issue (in the context of a given reference antecedent based on the same risk factor) jointly with modifiers of the RR's magnitude (propos. II – 2.2, 21–22).

Proposition III – 3.6: The RR at issue in the result of an etiognostic clinical study, most commonly by far, is *incidence-density ratio*, IDR, incidence density, ID, being the rate of an event's occurrence in the sense of the number of events per unit amount of population-time (ref.). Thus the *referent* of the result – of the empirical RR function – generally is a *study base* of the form of a particular aggregate of population-time (ref.), the one for which the IDR function was documented. (This contrasts with a series of person-moments as the base for a diagnostic study.)

   Reference: Miettinen OS. Estimability and estimation in case-referent studies. Am J Epidemiol 1976; 103: 30–6.

The Genesis of the Results of Etiognostic Clinical Studies

Proposition III – 3.7: The penultimate stage in the genesis of the empirical IDR function resulting from an etiognostic clinical study generally is the fitting of the logistic counterpart of the designed ID function to the 'data' on – actually the statistical variates' realizations in – two series: the *case series* ($Y = 1$) and *base/referent series* ($Y = 0$), the former representing all of the events at issue (typically inceptions of overt cases of the illness at issue) that occurred in the study base, the latter a fair sample of (the infinite number of person-moments constituting) the study base; and beyond this, the ultimate stage is the deduction of the empirical IDR function from the resulting logistic function (propos. II – 2.22).

Proposition III – 3.8: Those two series come about consequent to the adoption of a particular population as the study's *source population* and securing the first-stage case and base/referent series from the population-time of this population's course over a span of time, from the *source base* in this meaning. These first-stage series are reduced to the study's ultimate case and base/referent series from the actual *study population*, from a segment of its course over time, from the actual *study base*. Each person-moment in these reduced series represents the study object's domain and one of the histories in the causal contrast(s) of interest, both of these properties defined as of these person-moments in the series. The reduced, final series may also need to satisfy such practical criteria of belonging in the study base as were involved in its design (based on some of: place of residence, health insurance, language, being compos mentis, etc.).

Proposition III – 3.9: The source population may have a direct, primary definition; or it may be defined indirectly, secondary to the way in which the first-stage case series is identified – that is, as the *catchment population* of the means of case identification for this series. The catchment population is the entirety of those who, at a given time, are in the 'were-would' state of: were the illness event now to occur, it would be 'caught' into the first-stage case series. (Defined by this state, the catchment population is a dynamic one; i.e., it has turnover of membership.)

References:
1. Miettinen OS. Etiologic study vis-à-vis intervention study. Eur J Epidemiol 2010; 25: 671–5.
2. Miettinen OS. Theoretical Epidemiology. Principles of Occurrence Research in Medicine. New York: John Wiley & Sons, 1985; pp. 54–5.

The Quality of the Results of Etiognostic Clinical Studies

Proposition III – 3.10: What was said about the quality of the results of diagnostic clinical studies in respect to the role of the studies' object designs (propos. III – 2.11, etc.) generally applies, *mutatis mutandis*, to etiognostic clinical studies as well.

Proposition III – 3.11: The study result is free of *selection bias* if commitment to the source base was made without any basis for a hunch of what the IDR function (the magnitudes of its parameters' empirical values) for its associated study base might be, as distinct from these characterizing other potential selections for the study base. This aspect of validity is not inherently satisfied in etiognostic studies. (In diagnostic studies the truth about the presence/absence of the illness is determined only after the commitment to enrol the instance into the study series has been made; propos. III – 2.5, 16. An etiognostic study is inherently free of selection bias only if the source base is prospective in study time.)

Proposition III – 3.12: The study result is free of *documentation bias*, that is, assuredly descriptively valid for its referent – for the study base (which may or may not be free of selection bias) – if and only if: (a) the case series indeed is the entirety of cases that occurred in the study base (and does not include cases from outside the study base) or a random subset of this; (b) the base/referent series is a fair sample of the study base conditionally on the codeterminants (those other than the etiognostic one, incl. the factors by which the sampling of the source base was stratified) in the linear compound in the model for log(ID); (c) the 'facts' on these two series are correct; and (d) the fitting of the logistic counterpart of the designed log(ID) function was done correctly and was correctly translated into the corresponding IDR function (cf. propos. III – 3.7).

Proposition III – 3.13: The study result, if descriptively valid for the study domain (i.e., free of both selection bias and documentation bias), is subject to quantitative causal interpretation (i.e., free of confounding) if all extraneous determinants of (the magnitude of) the rate (ID) that were prone to have (or are known to have had) different distributions between the index and reference segments of the study base were suitably controlled (by suitable representation in the ID function) or were prevented from having the propensity to confound the study base (by suitable formation of the empirical version of the causal contrast).

The 'Cohort' and 'Trohoc' Fallacies in Epidemiologists' Etiologic Studies

Proposition III – 3.14: The true general nature – essence – of an etiologic study, whether for clinical or epidemiological (community-medicine) purposes, is that of the genesis of the study's result as set forth in propositions III – 3.7–9 above.

References:
1. Miettinen OS. Etiologic research: needed revisions of concepts and principles. Scand J Work, Envir & Health 1999; 6 (special issue): 484–90.
2. Miettinen OS. Commentary on the paper by Zhang et al. – Lack of evolution of epidemiologic methods and concepts. In: Morabia A (Editor). History of Epidemiologic Methods and Concepts. Basel: Birkhäuser Verlag, 2004.
3. Miettinen OS. Theoretical developments. In: Holland WW, Olsen J, Flurey C de V (Editors). The Development of Modern Epidemiology. Personal Reports of Those Who Were There. Oxford: Oxford University Press, 2007.
4. Miettinen OS. Etiologic study vis-à-vis intervention study. Eur J Epidemiol 2010; 25: 671–5.

Proposition III – 3.15: Proposition III – 3.14 and its references (above) notwithstanding, epidemiologists in their etiologic/etiogenetic studies continue to distinguish between 'cohort' and 'case-control' studies, or between 'cohort' and 'trohoc' studies. The 'trohoc' term is the (very fittingly evocative) heteropalindrome of 'cohort' (adduced by one of the two fathers of 'clinical epidemiology' in the contemporary usage of this term, the person whose name as the corresponding heteropalindrome would have been Navla Nietsnief).

Proposition III – 3.16: Contemporary epidemiologists have difficulty understanding – or in any case articulating (ref.) – the concept of cohort study in epidemiological research on the etiology/etiogenesis of an illness. Some so-called cohort studies – eminent ones such as the Framingham Heart Study and the Nurses' Health Study – have not been studies at all but, instead, programs of data collection for a database that provides for a wide variety of etiologic/etiogenetic studies, to be designed ad hoc and, quite possibly, as trohoc studies.

Reference: Porta M (Editor), Greenland S, Last JM (Associate Editors). A Dictionary of Epidemiology. A Handbook Sponsored by the I. E. A. Oxford: Oxford University Press, 2008.

Proposition III – 3.17: The true concept of *cohort study* (of the etiology of an illness) involves a study cohort – a population for which membership is defined by the event, at cohort T_0, of enrollment into it, starting as of this event and lasting forever thereafter. In such a study population, documented is *prospective* (post-T_0) occurrence of the illness in causal relation to retrospective (pre-T_0) divergence in the determinant (of the prospective rate of occurrence).

Proposition III – 3.18: The concept of cohort study as an etiologic/etiogenetic study is irrational. For a rational etiologic/etiogenetic study the object (in respect to the rate of occurrence) is *current* (at T_0 of etiologic/etiogenetic time) occurrence of the illness in causal relation to retrospective divergence in the determinant (cf. propos. III – 3.7–9 above). When a rationally construed etiologic/etiogenetic study is in progress, scientific time remains stalled at its T_0.

Proposition III – 3.19: The requisite remedy for the *cohort fallacy* (propos. III – 3.17–18 above) is to regard the cohort as the source population and the population-time of its follow-up as constituting the source base, to regard the prospectively identified cases as constituting the first-stage case series, and to draw the first-stage base series from this source base, etc. (cf. propos. III – 3.14 above).

Proposition III – 3.20: The concept of *case-control*/trohoc study is now 'officially' defined as "The observational epidemiological study of persons with the disease and a suitable control group of persons without the disease . . . comparing the diseased and nondiseased subjects with regard to how frequently the factor or attribute is present . . ."

> *Reference*: Porta M (Editor), Greenland S, Last JM (Associate Editors). A Dictionary of Epidemiology. A Handbook Sponsored by the I. E. A. Oxford: Oxford University Press, 2008.

Proposition III – 3.21: Different from the cohort study, the case-control/trohoc study is rational in the sense that it involves histories (in regard to the etio-logic/etiogenetic determinant of the rate's magnitude) as of the time of outcome; that is, as of etiologic/etiogenetic T_0 (cf. propos. III – 3.18).

Proposition III – 3.22: Different from the cohort study, the case-control/trohoc study is irrational in its failure to address and compare rates of the occurrence of the illness in a defined study base, and the consequent reversal of the comparison (trohoc!). So malformed is this conception of an etiologic/etiogenetic study that the alternative to causality – confounding (of the study base, to be controlled for the study result; propos. III – 3.13) – can never be understood from its vantage. (The reason for this is, principally, the absence of study base as an element in the concept of the case-control/trohoc study.)

Proposition III – 3.23: The requisite remedy for the *trohoc fallacy* (propos. III – 3.20, 22 above) begins with the necessary reconceptualization of what is involved in the structure of the study: not two groups of persons but two series of person-moments, one of them a case series and the other a non-case series. Next, the case series needs to be understood – like any case series in epidemiological practice or research – as being meaningful only insofar as it provides inputs to the derivation of rates in some defined population experience – and here, specifically, in a defined study base. Once this understanding has been achieved – and it is an utterly elementary one in the disciplines whose core concern is rates of morbid-ity (in human populations) – it should be obvious that the case – rate numerator –

series is to be coupled with a corresponding rate denominator series; that is, that the non-case series is to be construed as a sample of the study base. Once this much is understood, it remains to understand the elementary fact that numerators and denominators are inputs to division – here in the computation of quasi-rates – and not elements in a comparison. Step by step, understanding again leads to the structure of *the* etiologic/etiogenetic study (propos. III – 3.14 above).

Proposition III – 3.24: So long as epidemiologists have difficulty understanding their own, etiologic/etiogenetic research (in the service of community-level preventive medicine), they remain particularly unprepared to be authorities on quintessentially 'applied' clinical research – even though genuine competence in epidemiological research arguably is a prerequisite for gaining competence in such clinical research. (The latter type of research, ultimately addressing probabilities rather than rates per se, can be thought of as being meta-epidemiological.)

Reference: Miettinen OS. Epidemiology: quo vadis? Eur J Epidemiol 2004; 19: 713–8.

III – 4. INTRODUCTION INTO PROGNOSTIC CLINICAL RESEARCH

The Nature of the Results of Prognostic Clinical Studies

Proposition III – 4.1: While the GPF (gnostic probability function) results of diagnostic clinical studies are completely acausal, merely descriptive, in their intended interpretations, and while the corresponding results (IDR functions) from etiognostic studies – descriptive of experience like those of any empirical studies – are expressly designed for the purpose of causal inference, the results of prognostic clinical studies can, and now and in the future commonly should, have both of these qualities. For, the prospective course of the health of a modern doctor's client is, near-invariably, dependent – causally – on the choice of intervention (preventive or therapeutic) yet also of concern – acausally – conditionally on the choice of intervention.

Proposition III – 4.2: The PPF (prognostic probability function) results of prognostic clinical studies have both causal and acausal qualities when the determinants in them include the type of intervention along with the prognostic indicators (propos. II – 1.31), and when, in addition, the genesis of the result provides for causal inference about the probability estimate's dependence on the choice of intervention (cf. propos. II – 1.31).

Proposition III – 4.3: While the results of etiognostic studies address, as of the T_0 point of etiognostic time, current occurrence in causal relation to past/retrospective divergence in the etiogenetic determinant of the occurrence (propos. III – 3.18), the PPFs from intervention-prognostic studies address, as of the T_0 point of prognostic time, future/prospective occurrence in causal relation to prospective divergence in the intervention determinant, and in descriptive relation to the prognostic indicators' realizations at prognostic T_0 (cf. propos. II – 1.31).

Proposition III – 4.4: While the occurrence relations relevant to etiognosis translate into causality-oriented rate-ratios as functions of determinants of their magnitude (propos. III – 3.5), PPFs are based on absolute, proportion-type rates and on this

O. S. Miettinen, *Up from CLINICAL EPIDEMIOLOGY & EBM*,
DOI 10.1007/978-90-481-9501-5_12, © Springer Science+Business Media B.V. 2011

basis provide empirical values for intervention-conditional risks and for risk differences in the meaning of intervention-induced changes in the probabilities of the phenomena addressed in prognostication (propos. II – 1.31).

The Genesis of the Results of Prognostic Clinical Studies

Proposition III – 4.5: In the production of the results of prognostic clinical studies, the beginning is analogous to that in diagnostic studies (but not that in etiognostic ones): instances of the object function's domain (here one of prognostication) are tentatively identified in clinicians' practices, and the persons involved in these instances are, selectively, solicited for participation in the study; then, if informed consent is obtained, tentative enrollment into the study ensues.

Proposition III – 4.6: Upon tentative enrollment, analogously with diagnostic studies, *admissibility* into the study is assessed in detail, and with 'facts' whose likelihood of correctness about the domain criteria conforms to the requirements of research (and may exceed those of practice). These criteria include presence of the study indication for the interventions and freedom from contra-indications for these. Testing propensity to adhere to ('comply' with) assigned and agreed-upon medication use may be part of this assessment of admissibility when at issue is an experimental intervention-prognostic study – a 'clinical trial,' that is – on long-term pharmaco-interventions.

Proposition III – 4.7: With admissibility confirmed, *enrollment* of the person into the study cohort may – but need not – follow. Given enrollment, 'baseline' facts concerning the PPFs at issue – as to domain and prognostic indicators – are documented; and if the study is an intervention experiment, the particular intervention is now chosen – generally on the basis of random selection from among the compared options.

Proposition III – 4.8: In the course of a study subject's *follow-up*, documentation concerns the interventions and the health phenomena involved in the PPFs being studied. If the study is an intervention experiment, the assigned intervention is implemented if it requires healthcare personnel; otherwise the study subject's adherence to it is monitored and reinforced. In any case, the timing of and reason for the follow-up's termination is documented in a study pertaining to subacute or chronic prognosis (propos. II – 2.24).

Proposition III – 4.9: The study data are translated into the realizations of the statistical variates that are involved in the (predesigned) object PPFs (cf. propos. III – 2.6).

Proposition III – 4.10: Given the final aggregate of study data in the form of statistical variates' realizations, and given that at issue is acute prognosis, the predesigned logistic PPFs for the various types of outcome are fitted to these data (cf. propos. II – 2.25).

Proposition III – 4.11: When at issue is subacute or chronic prognosis about an event-type phenomenon of health, the data are to yield, first, an empirical function for the event's incidence density corresponding to the PPF at issue (as to domain and determinants); for, this is needed for the derivation of its corresponding function for cumulative incidence and, thereby, for prognostic probability (conditional on surviving extraneous causes of intercurrent death; cf. propos. II – 2.27, 28).

Proposition III – 4.12: In the production of a PPF for subacute or chronic prognosis, the data are to be approached – and the statistical variates modified – in the spirit of the etiognostic study (propos. III – 3.7–9), a particular variant of this. The series of cases of the event is identified from the database, and associated with each of these cases are somewhat modified variates: that the person-moment is associated with the event is indicated by $Y = 1$; the associated history of treatment (type, time lag since its implementation/initiation) is specified in terms of realizations for appropriate (in part newly defined) Xs; and the respective X realizations at prognostic T_0 are associated with each case. For this case series the referent – the study base – is understood to be constituted by the aggregate of the segments of person-time from prognostic T_0 to an endpoint that is the earliest one among the event at issue, death from an extraneous cause, loss to follow-up, and the study's 'common closing date.' A sample of this population-time aggregate is to be drawn for the *base series* and documented analogously with the case series (though with $Y = 0$); but a critically important feature of the 'etiogenetic study' for the purpose here is that the denominator series be a *representative* sample of the study base.

Reference:
1. Miettinen OS. Important concepts in epidemiology. In: Olsen J, Saracci R, Trichopoulos D (Editors). Teaching Epidemiology. Third edition. Oxford: Oxford University Press.
2. Hanley J, Miettinen OS. Fitting smooth-in-time prognostic risk functions via logistic regression. Internat J Biostat 2009; 5: 1–23.
3. Miettinen OS. Etiologic study vis-à-vis intervention study. Eur J Epidemiol 2010; 25: 671–5.

Proposition III – 4.13: Upon this modification of the usual type of dataset – variate realizations – from an intervention-prognostic (or merely descriptive-prognostic) study, the logistic counterpart of the predesigned object function $\log(\text{ID}) = L$ – the function $\log[\Pr(Y = 1)/\Pr(Y = 0)] = L$, that is – is fitted to the data; and the resulting $\hat{L}$ translates into the corresponding empirical ID function as follows:

$$ID = (b/B)\exp(\hat{L}),$$

where b is the size of the base series and B is the size of the base proper (in terms of amount of population-time; refs. in propos. III – 4.12 above). This ID, in turn, translates into an empirical probability function on the basis of its integral over the relevant interval of time (the prospective counterpart of the retrospective time in the rearranged data), according to proposition II – 2.27.

Proposition III – 4.14: When data of the usual type from a prognostic study are used to produce an empirical function for prospective prevalence/probability of a *state* of health, a study subject's potential for contribution to the study base – now a series (finite) of person-moments – does not end by the occurrence of the health events addressed in the propositions above (except if the event is death): the study series can be drawn – as a sample – from the entire population-time of follow-up. With the person-moments in this series documented in the manner of the series above (with $Y = 1$ if case present, 0 otherwise), the predesigned object of study – logistic prevalence/probability function (propos. II – 2.26) – is fitted to the data. The sampling's representativeness now is not a concern, but it must be independent of the presence/absence of the health state at issue.

The Quality of the Results of Prognostic Clinical Studies

Proposition III – 4.15: Focusing here on *intervention*-prognostic clinical studies and, specifically, *experimental* studies – 'clinical trials' – for subacute or chronic prognosis (for reasons of their relative commonality and importance), the quality of any given reported result of such a study again is a matter of the result's form for one – bearing on the study object's admissibility and relevance – and its empirical content for another – resulting from the design and execution of the methods of study, which accord a given degree of validity to the result.

Proposition III – 4.16: A good-quality result of an intervention-prognostic study does not address the effect(s) of a recommendation or intention to intervene; it addresses the effect(s) of an actual, defined type of intervention relative to an actual, defined alternative to this.

Proposition III – 4.17: For a good-quality result of an intervention-prognostic study, the contrasted interventions are defined – as algorithms – for the entire span of prospective time that is relevant in the context of the duration of the (follow-up and) outcome assessment, and as the exclusive interventions for that period. '*Usual care*' is not a defined algorithm of intervention and, thus, not an admissible element in the object of an intervention-prognostic study; a good-quality result of an intervention-prognostic study does not contrast closely defined (still unusual) care with essentially undefined (melange of at present usual) care.

Proposition III – 4.18: For a good-quality result of an intervention-prognostic study, each of the contrasted interventions is a candidate for becoming the intervention-of-choice, defined as an algorithm for application, as the only intervention, throughout the time horizon of the prognosis.

Proposition III – 4.19: A good-quality result of an intervention-prognostic study addresses a well-defined intervention contrast (propos. III – 4.16–18 above) in terms of intervention-conditional, proportion-type rates (empirical) as functions (descriptive) of prognostic indicators (at prognostic T_0), and for subacute or chronic prognosis also of prognostic time, with a difference in these rates between the interventions subject to causal interpretation (cf. propos. II – 1.31).

References: See proposition III – 4.12.

Proposition III – 4.20: For a good-quality result of an intervention-prognostic study, the provisions for *validity*, for quantitative causal interpretability include: (a) (a good process of) randomization of intervention assignments (ref.), (b) close adherence to the randomly assigned intervention (propos. III – 4.16), (c) essential freedom – based on 'blinding,' if necessary – from prospective (post-randomization) confounding (propos. III – 4.17), and (d) essentially error-free documentation of both the contrasted interventions and the outcome at issue.

Reference: Miettinen OS. The need for randomization in the study of intended effects. Statist Med 1983; 2: 267–71.

Proposition III – 4.21: Apart from concern to assure validity of the results of an intervention-prognostic study, the methods design of such a study involves concern to maximize the study's *efficiency*, as an added dimension of quality of the study (though not of its result; cf. propos. III – 1.6). A major determinant of the study's efficiency is the study population's distribution according to both the prognostic indicators (cf. propos. III – 2.19) and by the type of intervention. As for the latter, the efficiency-optimal allocations are inversely proportional to the respective unit costs of intervention-cum-follow-up.

On Guidelines for Reporting on Clinical Trials

Proposition III – 4.22: Different from the practice of clinical medicine, the conduct of clinical research and reporting on its resulting evidence should be understood not to be subject to leaders' authority. Scientific communities should be understood to be constitutionally egalitarian, ones in which relevant for cogency of ideas is only the reasoning and evidence whence the ideas derive, never their presenters' standings in some hierarchies. *No person, committee, or whatever entity in a position of power should use the power to dictate what constitutes good research or, even, good reporting on research.* For, "Science flourishes best when it [is] unconstrained by preconceived notions of what science ought to be" (ref. 1). (Accordingly, this course was a matter of mere propositions for the students to individually weigh and consider; propos. I – 1.1.) The principle relevant to this is implicit in this theological precept: "And even as each one of you stands alone in God's knowledge, so must each one of you be alone in his knowledge of God and in his understanding of the earth" (ref. 2). Each genuine clinical scholar must ultimately be alone in his/her understanding of quintessentially 'applied' clinical research.

References:
1. Dyson F. The Scientist as a Rebel. New York: New York Review of Books, 2008; p. 17.
2. Gibran K. The Prophet. New York: Alfred A. Knopf, 1970; p. 57.

Proposition III – 4.23: It should be understood that a study report's acceptability for publication is properly determined by *peer review* (refs.) alone, unconstrained by any editorial 'requirements' for (i.e., editors' preconceived notions of) what constitutes good-quality 'reporting' on (the evidence from) a study, especially when a requirement on 'reporting' actually stipulates what the nature of the study ought to be. (Editors do have the power to dictate requirements to researchers, but use of it impedes the progress of science; cf. propos. III – 4.22 above.)

References:
1. Godlee F, Jefferson T (Editors). Peer Review in Health Sciences. Second edition. London: BMJ Books, 2003.
2. Lamont M. How Professors Think. Inside the Curious World of Academic Judgment. Cambridge (MA): Harvard University Press, 2009.

Proposition III – 4.24: "Can medical journals lead or must they follow" is a chapter heading in an important book (ref.). It explains that while their existence has been questioned, there arguably are leadership roles for "medical journals," meaning for their editors. Very notably, however, this eminent source does not present development of 'requirements' for acceptability of research reports as one of the possible leadership roles for editors of medical journals.

Reference: Smith R. The Trouble with Medical Journals. Glasgow: Royal Society of Medicine Limited, 2006; chapter 4.

Proposition III – 4.25: Rather than under editors' 'requirements,' clinical researchers and peer reviewers of their reports should function under the dominion of the theory of clinical research, the dictates of reason codified in this; and where these dictates remain incompletely developed or understood or agreed upon, resolution of the variance of opinions should be sought in the usual way of science – by public discourse in the relevant scientific community.

Proposition III – 4.26: Editors of medical journals at large, quite unjustifiably, act as authorities on the theory of clinical research. Thus, as for "reporting" on clinical trials, they declare (ref. 1) that investigators "should refer to the CONSORT statement [ref. 2]."

References:
1. International Committee of Medical Journal Editors. Uniform requirements for manuscripts submitted to medical journals: writing and editing for biomedical publication. (Updated October 2008.) www.icmje.org.
2. CONSORT Statement 2001 – Checklist. Items to include when reporting on a randomized clinical trial. www.consort-statement.org.

Proposition III – 4.27: The CONSORT statement the teacher of this course takes to be at variance with the dictates of reason for (i.e., the theory of) clinical *research itself* (apart from its reporting), in a number of ways. Some examples concerning "reports" on clinical trials may suffice to illustrate this:

1. The report is to include, the editors' "statement" says, a description of "How sample size was determined," while also including "the estimated effect size and its precision (e.g., 95% confidence interval)." While the editors' writing in this calls for editing, the main point of note here – one of principle of research per se – is this proposition: When the result of a study, with a specified precision, is at hand, it does not matter for the result's evidentiary meaning how that degree of precision came about (i.e., how the study's size and efficiency got to be what they were); and just as irrelevant to the result's evidentiary burden is, how well – or poorly – the actually attained degree of precision could be surmised by means of whatever "sample size determination" was carried out in designing the trial, this "determination" being an exercise in mere pseudo-optimization of study size (ref.).

   Reference: Miettinen OS. Theoretical Epidemiology. Principles of Occurrence Research in Medicine. New York: John Wiley & Sons, 1985; p. 62.

2. The report is to include, the editors' "statement" says, "explanation of any interim analyses and stopping rules." But here's a relevant proposition: Contrary to the claims of traditional, 'frequentist' statisticians, these aspects of a clinical trial have no bearing on the evidentiary significance of the results from the vantage of modern, Bayesian statistics (nor as a matter of common intuition).

   References:
   1. Cornfield J. Sequential trial, sequential analysis and the likelihood principle. Am Statist 1966; 20: 18–23.
   2. Berry DA. Interim analyses in clinical trials: classical vs. Bayesian approaches. Statist Med 1985; 4: 521–5.

3. The report is to have, the editors' "statement" says, "Clearly specified primary and secondary outcome measures," and as to the "analyses," and indication of "those pre-specified and those exploratory." To be weighed and considered here is this: "the data provide evidence only … and this evidence is altogether independent of the investigator's … mind-set before becoming aware of that evidence. … Finally, it is good to bear in mind that … [n]one of the formulations involved [in 'frequentist' statistics] address the history of the mind-set of the investigator in any way. Bayesian statistics does address subjective credibility of hypotheses, and its formulations for inference make no distinction between prior hypotheses and those suggested by the data" (ref.). The meaning of 'prior' in this context is: prior to the change resulting from the evidence at issue.

   Reference: Miettinen OS. Theoretical Epidemiology. Principles of Occurrence Research in Medicine. New York: John Wiley & Sons, 1985; p. 114.

4. The report is to include, the editors' "statement" says, addressing "multiplicity by reporting any other analyses performed" and discussion which takes into account "the dangers associated with multiplicity of analyses and outcomes." This, however, is yet another one of those 'frequentist' doctrines without a foundation in frequentist statistics (other than the here irrelevant theory of multiple contrasts in the context of a single, multicategory, nominal-scale determinant).

References:
1. Miettinen OS. Theoretical Epidemiology. Principles of Occurrence Research in Medicine. New York: John Wiley & Sons, 1985; p. 115.
2. Miettinen OS. Up from 'false positives' in genetic – and other – epidemiology. Eur J Epidemiol 2009; 24: 1–5.

5. The report is to include, the editors' "statement" says, discussion of "Generalizability (external validity) of the trial findings." At issue presumably is "The degree to which results of a study may apply, be relevant, or be generalized to populations or groups that did not participate in the study," a study being said to be "externally valid, or generalizable, if it allows unbiased inferences regarding some other specific target population beyond the subjects in the study" (ref.). Now the need is to weigh and consider that this is the concept of validity in the sample-to-population generalizations that are germane to sample surveys, while in empirical science the 'generalization' – inference, really – is from the empirical (and thereby particularistic, spatio-temporally specific) to the abstract (placeless and timeless) domain of the object of study; that in science there are no "target populations" and, hence, no sample surveys; that the concepts of 'generalizability' and 'external validity' in respect to clinical trials – as well as other types of clinical research – ought to be replaced, simply, by validity in reference to the domain (abstract) of the object of study (cf. propos. III – 1.6, 4.20).

Reference: Porta M (Editor), Greenland S, Last JM (Associate Editors). A Dictionary of Epidemiology. A Handbook Sponsored by the I. E. A. Fifth edition. Oxford: Oxford University Press, 2008.

6. The report is to include, the editors' "statement" says, the authors' "Interpretation of the results." But, who cares? Here is something to weigh and consider: The authors may not even be members of the relevant community of scientists; and if they are, their inference from the evidence they themselves have produced – results in conjunction with their genesis (propos. III – 1.5) – presumably is quite atypical of that of the relevant community of scientists at large; and insofar as at issue is evidence from an original study, inference from it alone rarely is a concern for that scientific community, even.

Proposition III – 4.28: The ultimate concern of editors of medical journals properly is not the quality of the manuscripts that are submitted for publication but the *quality of the actually published reports*, including *as a subset* of all of the reports that have been submitted for publication. Given a worthy object of study and valid methodology of studying it, peer reviewers of the study report's manuscript together with the editor(s) can forge from it a publishable report. But critical for the quality of the aggregate of actually published study reports on whatever clinically relevant object of study is *acceptance for publication independently of the results* (ref.). This means that editors should see to it that manuscripts go to peer reviewers with a view to assuring that their recommendations for acceptance/rejection will be independent of the results, that is, without editors' results-based screening and without the results; and that the editorial decision about acceptance for publication also is independent of the results, that is, taken prior to knowing anything about the results. By the same token, "A truly responsible investigator collects only such data as he/she is also determined to submit for publication, regardless of what those data seem to imply" (ref.).

Reference: Miettinen OS. Theoretical Epidemiology. Principles of Occurrence Research in Medicine. New York: John Wiley & Sons, 1985; p. 67.

Proposition III – 4.29: The CONSORT statement/checklist of "items to include when reporting a randomized trial" should be replaced by a *single editorial requirement*, concerning what *not* to include: A manuscript qualifies for being considered for publication only if submitted without any results and without any content based on these.

Proposition III – 4.30: *Good editorial policies* would not stipulate even the section titles for a report on quintessentially 'applied' (or other) clinical research – at least not in a way that is different between the report's summary and the report proper (ref.). That editors' understanding of the issues is incomplete is evinced, for example, by the fact that the stipulations for the summary's/abstract's section titles/topics (in the context of a given genre of research) varies substantially among journals, and are regularly different from the counterparts of these in the report proper – and that both of these are without a section for the object(s) of study in front of that for the method(s).

Reference: Miettinen OS. Evidence in medicine: invited commentary. CMAJ 1998; 158: 215–21.

Proposition III – 4.31: While editors of medical journals should not presume to be experts on quintessentially 'applied' medical research and in any case not dictate the terms of such research, not even of the ultimate reporting on it (propos. III – 4.22), they should be more competent in, and/or serious about, actual *editing* than they are at present – as will be evident in section IV – 2. The important lapses in this are not matters of style but of substance: poor writing, even in the most eminent journals is, commonly, misleading even to fellow researchers, to say nothing about non-researcher colleagues or science-writers for the general public.

PART IV
CONTEMPORARY REALITIES
IN CLINICAL RESEARCH

IV – 1. ON EBM GUIDELINES FOR ASSESSMENT OF EVIDENCE
         EBM Precepts Overall: Their Assessments
         EBM Precepts re Diagnostic Research: Their Assessments
         EBM Precepts re Prognostic Research: Their Assessments

IV – 2. SOME EXAMPLE STUDIES: THEIR ASSESSMENTS
         Examples in the Teachings about EBM
         Diagnostic Research: A Paradigmatic Study
         Diagnostic Research: Paradigm Lost, Example Series I
         Diagnostic Research: Paradigm Lost, Example Series II
         Etiognostic Research: Clinical Examples
         Prognostic Research: Clinical Examples
         Screening Research: Epidemiological Examples
         Screening Research: A Clinical Program

IV – 1. ON EBM GUIDELINES FOR ASSESSMENT OF EVIDENCE

EBM Precepts Overall: Their Assessments

Proposition IV – 1.1: Even though the body of EBM doctrines has evolved from a highly objectionable seminal idea (propos. I – 5.5–8), it deserves some further examination because of the prevalent touting of EBM at present, even if, typically, only by medical academics who never have studied those precepts (cf. academic 'Marxists' in the 1960s and '70s).

Proposition IV – 1.2: The most authentic codification of the EBM doctrines is, arguably at least, the most recent textbook of it in which D. L. Sackett still was the lead author. After all, it was he who had the seminal inspiration: "it dawned on him that epidemiology and biostatistics could be made as relevant to clinical medicine as his research into the tubular transport of amino acids," and this seminal idea of his led to 'clinical epidemiology' as the theoretical foundation of EBM (propos. I – 5.3; ref.).

> *Reference*: Sackett DL, Straus SE, Richardson WS, et alii. Evidence-Based Medicine. How to Practice and Teach EBM. Second edition. Edinburgh: Churchill Livingstone, 2000; p. ix.

Proposition IV – 1.3: According to the most authentic source (above), "Evidence-based medicine (EBM) is the integration of best research evidence with clinical expertise and patient values. . . . When these three elements are integrated, clinicians and patients form an alliance which optimizes clinical outcomes and quality of life" (p. 1). The nature of that "alliance" and "integration" are taken to be self-evident and are thus left without any explication. But, whatever may be the "alliance" and "integration" in the essence of EBM (insurmountable hermeneutic challenges abound in the doctrines of EBM; cf. propos. II – 2.6), this pair of ingredients cannot conceivably be so magical that it, by its very nature, "optimizes clinical outcomes and quality of life." For neither party to that alliance really knows what the outcomes and quality of life will be; and scarcely are they assured to be optimal (even probabilistically) when the patient's potential alliances with other EBM clinicians – with their different assessments of evidence and different levels of relevant clinical expertise – would tend to mean different outcomes and different qualities of life.

O. S. Miettinen, *Up from CLINICAL EPIDEMIOLOGY & EBM*,
DOI 10.1007/978-90-481-9501-5_13, © Springer Science+Business Media B.V. 2011

Proposition IV – 1.4: "The full-blown practice of EBM," the relevant source (propos. IV – 1.2) says, "comprises five steps." "Step 1" in this is "converting the need for information … into an answerable question" (p. 3). But, given a client presentation in a clinician's practice, there never is an inherent and recognizable need for extrinsic "information." The general need truly is to know what facts to ascertain and how to convert the ascertained facts into gnosis (propos. II – 1.13–14, etc.). Where the doctor doesn't know these things, (s)he is to convert the ignorance into the *relevant* question(s), whether answerable or not.

Proposition IV – 1.5: While a practitioner of EBM, when not knowing what to do or think, takes "Step 2 – tracking down the best evidence with which to answer the question" (p. 3), a practitioner of rational medicine, which is KBM (knowledge-based medicine; propos. II – 2.5), consults a source of the needed knowledge, founded on all of the relevant evidence and whatever else bears on it (propos. I – 5.11). So long as a relevant expert system remains unavailable, rational clinicians see good justification for seeking the knowledge "from direct contact with local experts" or from "writings of international experts," dismissing the founding doctrine of the EBM cult (propos. I – 5.5).

Proposition IV – 1.6: While a practitioner of EBM, when presuming to have tracked down "the best evidence," takes "Step 3 – critically appraising that evidence for …" (p. 4), a genuine professional does not accord highest credibility to his/her own opinion about what the "best" evidence is and what to make of it (this as a substitute for the genuine role of evidence in science; propos. III – 1.14–16). (S)he simply defers to the expert(s) (s)he consults (propos. I – 5.14).

Proposition IV – 1.7: While a practitioner of EBM presumes to be able to next take "Step 4" as a matter of "integrating the critical appraisal with ... [the] patient's unique biology," etc. (p. 4), a practitioner of KBM accords no virtue into thinking of the patient's "biology" as being "unique." Instead, (s)he thinks of the instance at hand as representing a definable type of recurrent challenge and takes the accent on uniqueness to be tantamount to denying the possibility of KBM (propos. II – 2.6) and thereby of clinical professionalism.

Proposition IV – 1.8: The fifth and final "step" in the "full-blown practice of EBM" the relevant source specifies as "evaluating our effectiveness and efficiency in executing steps 1–4 and seeking ways to improve them both next time" (p. 4). But, whatever may be the effectiveness of this "independent assessments of evidence" (propos. I – 5.5) and this appraisal's (unjustifiable) "transformation" into "direct clinical action" (propos. I – 5.12–13), the inefficiency of each practitioner tracking down and appraising the same evidence is so obviously enormous as being solely sufficient for judging the overall body of EBM doctrines to be, well, absurd. Rational ideas about the pursuit of improved efficiency of healthcare are very different (propos. III – 3.1–3).

EBM Precepts re Diagnostic Research: Their Assessments

Proposition IV – 1.9: To the 'clinical epidemiologists' who have been and are the leaders of the EBM movement (propos. IV – 1.2), diagnostic research, in general, produces "evidence about the accuracy of a diagnostic test" (p. 67), a test being "an item of the history or physical examination, a blood test, etc." (p. 68). The evidence results from the "test's" "comparison with a reference ('gold') standard of diagnosis" (p. 68). The evidence addresses the "ability of this test to accurately distinguish patients who do and don't have a specific disorder" (p. 72), the measures of this ability being "sensitivity, specificity, and likelihood ratios" (p. 72). Given the result from one of the "tests," the corresponding likelihood ratio – the same regardless of the pre-test facts – is used to make the transition from the pre-test probability to the corresponding post-test probability (according to Bayes' theorem; p. 73). As for the pre-test probability, "We've used five different sources for this vital information: clinical experience, regional or national prevalence statistics, practice databases, the original report we used for deciding on the accuracy and importance of the test, and studies devoted specifically to determining pre-test probabilities" (p. 82). This aggregate of ideas is seriously flawed, starting with the failure to appreciate that, in all of science, *accuracy* is a feature of measurement – quantification – only, and not of classification (on a nominal or ordinal scale; ref.).

> *Reference*: Olesko K. Precision and accuracy. In: Heilbron JL (Editor-in-Chief). The Oxford Companion to the History of Modern Science. Oxford: Oxford University Press, 2003; pp. 672–3.

Proposition IV – 1.10: Regarding a study on the "accuracy of a diagnostic test," one of the validity questions is said to be, "Was [the "accuracy"] evaluated in an appropriate spectrum of patients (like those in whom we use it in practice)?" (p. 68). And as for "critically appraising a report about pre-test probabilities of disease," one of the validity questions is said to be, "Did the study patients represent the full spectrum of those who present with this clinical problem?" (p. 83). These ideas about validity imply understanding that both the measures of the "accuracy of a diagnostic test" and the pre-test probability lack universality of value; that they are not singular in value. Given this understanding, it should be understood to be irrational to study the (ill-definable) typical values of these measures and probabilities and to apply these typical values in the practice of diagnostic probability-setting. Relevant distinction-making is in the essence of rational diagnosis (propos. II – 2.5).

Proposition IV – 1.11: Another one of those EBM questions concerning the validity of a study on the "accuracy" of a "test" – ascertainment of the presence/absence of a symptom, say – is this: "Was the reference standard applied regardless of the test result?" (p. 68). Indeed, the empirical values of those measures of the diagnostic "accuracy" of whatever potential manifestation of the illness at issue obviously are distorted if this very manifestation of the illness – or any correlate of this, even – has a role in the instances' becoming entries into the study series. But, elimination of this influence – that is, formation of a study series consisting of instances in all of which the presence/absence of the illness at issue got to be determined without any role for the manifestational profile in the prompting of this determination – is practically unimaginable. After all, patients generally present themselves for the pursuit of diagnosis because of the manifestations of their illness.

Proposition IV – 1.12: Once adopted is the (strange) view that at issue in diagnostic research always is a "test" in the inclusive sense of "an item of the history or physical examination, a blood test, etc." (propos. IV – 1.9) and that the pre-test probability before the very first one of these "tests" could be "regional or national prevalence statistics" (*sic*), the question arises, Why not think of *all* of the diagnostic probabilities in terms of the prevalence of the illness conditional on the diagnostic profile at hand, with no pre-test versus post-test duality in this? Moreover, and more specifically, Why not think of the diagnostic probability – based on prevalence – as a *joint function* of the diagnostic indicators in the logistic-regression framework (propos. II – 2.14), which inherently accounts for the (partial) redundancies among the sequential "tests" – the intercorrelatedness of their results – thus avoiding the overinterpretation of the discriminating significance of the diagnostic profile, which is a major problem with sequential updating of the diagnostic probability on the basis of the "tests'" respective measures of "accuracy" (cf. propos. II – 1.23).

Proposition IV – 1.13: An extensive, uncritical account of the 'clinical epidemiology'-and-EBM culture of diagnostic research is given in a recent textbook (ref.).

> *Reference*: Knottnerus JA, Buntinx F. The Evidence Base of Clinical Diagnosis. Theory and Methods of Diagnostic Research. Second edition. Chichester (UK): BMJ Books / Wiley-Blackwell, 2008.

EBM Precepts re Prognostic Research: Their Assessments

Proposition IV – 1.14: The leading 'clinical epidemiologists' who in their book on EBM (propos. IV – 1.2) teach doctors about clinical research with a view to the practice of EBM, distinguish between "evidence about prognosis" (p. 95) and "evidence about therapy" (p. 105). Now, diagnostic research as it has been addressed in the foregoing does not produce evidence *about* diagnosis but, instead, *for* (the development of the knowledge-base for) diagnosis; rather than about diagnosis, the evidence from diagnostic research, when properly construed, is about profile-conditional prevalence of the illness at issue, about the way in which this is a joint function of the diagnostic indicators that have been accounted for (propos. III – 2.1–2). Analogously, prognostic research, when properly construed, produces evidence not about prognosis but for prognosis; and what it is about is prospective incidence/prevalence of a health event/state, conditionally on the prognostic profile and the choice of intervention (propos. III – 4.1–4). The idea that prognosis concerning the future course of a modern person's health – and the evidence for it – can generally be meaningfully addressed separately from, and thus without specificity to, intervention is not realistic (nor is prophylactic intervention "therapy").

Proposition IV – 1.15: The first question about the validity of whatever evidence "about prognosis" is said to be, "Was a defined, representative sample of patients assembled at a common (usually early) point in the course of the disease?" (p. 95), as "Ideally, the prognosis study we find would include the entire population of the patients who ever lived who developed the disease, studied from the instant of its onset" (p. 96). But to be at all realistic, it should be understood that clinical prognosis is usually set in the course of the patient's already-overt (clinically manifest) case of an illness, and repeatedly reset with updatings of the prognostic profile (incl. in respect to history of interventions) and of the contemplated prospective treatments. A well-designed prognostic probability function, PPF, fitted to the data from a non-representative melange of cases of the illness – including in respect to the stage of its progression – addresses, suitably, the multitude of the situations that are involved in the context of any given illness, making the requisite distinctions (propos. II – 2.3, 24–29).

Proposition IV – 1.16: As those leaders' teaching turns to intervention research (pp. 105 ff), they overlook the important fact that an intervention trial produces evidence about intervention-conditional future course of health, relevant to the knowledge-base of intervention-conditional prognosis, so that at issue should be understood to be prognostic research; and that more relevant to the decision about a particular intervention than the difference(s) between the prognoses conditional on this intervention and its alternative(s) are the intervention-conditional prognoses as such (propos. II – 1.31); and that those intervention-conditional prognoses need to be specific to the person's prognostic profile and commonly also to periods/points of prognostic time (propos. II – 2.24–29, IV – 1.15 above). In this context, as in others, those leaders of 'clinical epidemiology' and EBM overlook the problem of multiplicities in the requisite knowledge-base of clinical medicine (propos. II – 2.1–6). To teach every clinician to (presume to) understand clinical research, they trivialize the research by not addressing suitably distinctions-making PPFs (propos. IV – 1.15 above), just as the diagnostic counterparts of these have remained alien to them.

IV – 2. SOME EXAMPLE STUDIES: THEIR ASSESSMENTS

Examples in the Teachings about EBM

Each of the three discussion groups formed from the students in this course selected three published articles on clinical research for review in class, and one of the nine articles was:

> Grover SA, Barkun AN, Sackett DL. Does this patient have splenomegaly? JAMA 1993; 270: 2218–21.

This article appeared in the journal's section entitled "The Rational Clinical Examination."

While not a report on clinical research – original or derivative – conducted by the authors, this article nevertheless is instructive in the context here. For it gives a strong indication of how leading 'clinical epidemiologists' think about evidence-based diagnosis in the practice of EBM.

The titles of the article's successive sections are these: Three patients, Why examine the spleen?, Anatomic landmarks and spleen size, How large is normal spleen?, The consequences of splenomegaly for the clinical examination, How to examine for splenomegaly (with subsections Inspection, Percussion, and Palpation), Precision of the signs for splenomegaly, Accuracy of the signs for splenomegaly, Is splenomegaly ever normal?, The bottom line (Table 3), and Back to the bedside. That "bottom line" is the authors' conclusion about the attainability of practically definitive (rule-in or rule-out) diagnosis about splenomegaly by percussion and palpation of the spleen.

At least two of the three (hypothetical) patients are, for different reasons, examined "for splenomegaly" by percussion and palpation. The article implies that the results of these examinations (positive/negative) are to be translated into probability of splenomegaly in the context of a given level of "clinical suspicion" before these examinations, using these examinations' respective "accuracies" in terms of "sensitivity" (probability of positive finding, given presence of splenomegaly) and "specificity" (probability of negative finding, given absence of splenomegaly).

The article does not say anything about the way in which the level of "clinical suspicion" before the percussion and palpation of the spleen is to be set, nor about the way in which this "prior probability or disease prevalence" is to be translated into the corresponding post-examination probability of the patient having a case of splenomegaly.

The article does, however, present tables citing the results of studies on the respective "accuracies" ("sensitivity" and "specificity") of the spleen's percussion and palpation for the diagnosis about splenomegaly. The article also makes the point that one study had addressed the palpation's "discriminating ability" (as to presence/absence of splenomegaly) separately according as the percussion finding was positive or negative, and that, evidently, the palpation's "discriminating ability" was essentially confined to those with a positive finding from the percussion; but it says nothing about the corresponding – presumably different – values of the palpation's "sensitivity"/"specificity" depending on what the percussion finding is (cf. propos. II – 2.18).

While focusing on diagnosis about splenomegaly, this article seems to have been designed to teach the readers in a more general way that, while the items in a patient's clinical examination obviously have to be carefully thought out and executed, it also is necessary to know about the "accuracy" ("sensitivity" and "specificity") of each of these – on the basis of published reports on clinical research. And indeed, according to textbooks of 'clinical epidemiology' (propos. I – 3.1), if P' is the diagnostic probability prior to the incorporation of a given (manifestational) item into the diagnostic profile, the corresponding probability after its incorporation, P'', is implicit in the relation $P''/(1-P'') = [(1-P')/P'] \times LR$, where LR is the datum's likelihood ratio, specifically $LR_+ =$ 'sensitivity' / (1 – 'specificity') for a positive finding or $LR_- = (1 -$ 'sensitivity') / 'specificity' for a negative finding, and where both "sensitivity" and "specificity" are treated (unjustifiably; propos. II – 2.18) as though independent of the pre-test profile.

Now, the first patient's presentation – "an elderly woman who complains about easy fatigability" – in conjunction with the observation (in clinical examination) that "her conjunctivae and nail beds are pale" is said to make the clinician "suspect that she is anemic due to gastrointestinal blood loss," but to nevertheless elect (inexplicably) to first focus on "a lymphoproliferative disorder" as an element in the differential-diagnosis set. The second patient's presentation – "a college student with failing appetite, ability to concentrate, energy, and grades" – is said to make the diagnostician think that the student is depressed, but to want first to "rule out infectious mononucleosis."

If we take it that, for some reason (quite inapparent), diagnosis about "a lymphoproliferative disorder" or infectious mononucleosis is to be pursued clinically, before definitive diagnosis by laboratory examination of peripheral blood and perhaps also bone marrow for the former and blood-smear and serological tests for the latter, the point of departure is to be the proper, relevant orientational question (rather than an "answerable" one; propos. IV – 1.4).

The relevant orientational question is not, in either one of these cases, Does this patient have splenomegaly? Instead, the questions, in respect to each of those illnesses of differential-diagnostic concern – splenomegaly not being among them – are: What set of clinical facts are to be ascertained?; and given the ascertained set of facts, What is the proportion of instances like this in general such that the illness at issue is present? (Propos. II – 1.14.)

The facts from physical examination that bear on diagnosis about chronic lymphocytic leukemia do not derive from percussion and palpation of the spleen alone; those of the liver also matter, along with findings concerning not only pallor but also generalized skin lesions and generalized lymphadenopathy; and all of these bear on the diagnosis in conjunction with facts from history-taking – and, notably, without the facts' interpretation in respect to presence/absence of hepatomegaly (secondary to CLL). Much of this applies to clinical diagnosis about infectious mononucleosis just the same.

The notion that the probability of a diagnostically relevant binary datum turning out to be positive in the presence of splenomegaly – the item's 'sensitivity' to splenomegaly, that is – is singular in value is seriously incorrect. Palpation of the spleen presumably gives a positive finding more commonly in those with positive finding from percussion for splenomegaly (cf. above), as the latter finding suggests a relatively high degree of splenomegaly insofar as there is any.

Similarly, both of these examinations are more likely to give a positive result in the presence of splenomegaly if the probability of splenomegaly before either of them is exceptionally high, as this high prior probability generally results (in part at least) from (strongly) positive findings on (many of) the other manifestational indicators, and this again points to a relatively high degree of splenomegaly insofar as any of this is present. (Grover et alii entertained prior probabilities as high as 90%, highly unjustifiable by the patient vignettes presented.)

Addressing the diagnostic profile in terms of successive application of item-specific 'univariate' likelihood ratios – whether based on 'sensitivity' and 'specificity' or whatever – is akin to deriving the coefficients for a multivariate logistic probability function (propos. II – 2.14) from the fittings of the corresponding set of univariate models: the indicators' mutual redundancies/correlations are falsely treated as though nil (propos. II – 1.23), with the consequence of potentially very serious overestimation of the profiles' discriminating informativeness. The findings from the spleen's percussion and palpation are correlated, mutually redundant (to an appreciable extent).

Accuracy, properly construed, characterizes quantitative facts only, the measurements/quantifications represented by them (propos. IV – 1.9). The spleen's percussion and palpation, each with a binary result, do not represent quantification of anything; and, for this reason, they are not characterized by their respective accuracies. Measurement of the size of the spleen (cinti- or sonographically) for diagnosis about CLL, say, would be characterized by its accuracy – but only in terms of the general distribution of the degree of agreement of its result with the true size of the spleen and, very notably, without any regard for the presence/absence of the CLL or whatever other object of the diagnosis. Moreover, test accuracy in this genuine meaning of the term *can* be regarded as independent of the pre-test probability of the illness at issue being present.

The finding from the spleen's percussion or palpation, it should be understood, is positive or negative merely in respect to what is heard or felt, and not in respect to (inference about) presence/absence of splenomegaly. A positive finding therefore is not 'true-positive' or 'false-positive' according as splenomegaly is or is not present but, instead, according as the finding as such is correct or incorrect. (On a nominal scale there is no degree of a finding's/datum's agreement with the truth that it ideally would represent; there either is, or isn't, agreement.)

Insofar as one does (unjustifiably) regard 'sensitivity' and 'specificity' as measures of a diagnostic indicator's 'accuracy,' they must be thought of as parameters characterizing a binary indicator's distributions in general, in the abstract, not in the experience of a particular study. It thus is incorrect to say, for example, that one method of palpation "exhibited a significantly (P < .05) higher sensitivity (82% vs 59%) but lower specificity (83% vs 94%)" than the other. Rather than the (falsely-presumed universal) values of the 'sensitivity' and 'specificity' parameters, those percentages are merely their corresponding empirical frequencies in the studies at issue.

'Disease' and 'diagnosis' are not synonyms in scholarly clinical jargon: the former (propos. II – 1.6) has to do with the soma of the patient, the latter (propos. II – 1.13) with the mind of the doctor. It thus is incorrect to refer to the diseases causing splenomegaly as "these diagnoses," and to use interchangeably the terms "rule in splenomegaly" and "rule in the diagnosis of splenomegaly."

Very notably, the authors of the article say nothing at all about critical evaluation of the studies producing the empirical measures of "accuracy" it displays, that is, about "step 3" in the "full-blown practice of EBM" (propos. IV – 1.6).

Further examples of the use of examples in the teachings about EBM are reasonably sought from:

> Sackett DL, Straus SE, Richardson NS, et alii. Evidence-Based Medicine. How to Practice and Teach EBM. Second edition. Edinburgh: Churchill Livingstone, 2000.

The relevant chapters to explore are those entitled Diagnosis and Screening, Prognosis, Therapy, and Harm. The respective numbers of examples of the use of evidence are: one, none, none, and none; and that one, even, does not really qualify as an example of the evaluation and use of evidence in the practice of EBM, ultimately as a matter of "the integration of best research evidence with clinical expertise and patient values" (propos. IV – 1.3).

That single example concerns a hypothetical *patient with anemia*, in the case of which we "think that the probability that she has iron deficiency anemia is 50%" (p. 72). Nothing is said about the genesis of the thought that the probability for iron deficiency anemia is 50%, notably as to the role of "critically appraising" evidence in it (propos. IV – 1.6). The chapter does have a subsection entitled Can We Generate a Clinically Sensible Estimate of Our Patient's Pre-test Probability?, said to be "a key topic" (p. 82) concerning which the authors have used "five different sources for this vital information" (p. 82; cf. propos. IV – 1.9). The pre-test probability of the lady with anemia, addressed above, does not come up as an example of the use of these sources, nor is there any other example of their use.

The example concerns the use of the *serum ferritin* test – as one of the "predictors, not explainers, of diagnoses" (p. 71). The decision to perform this assessment also seems to be taken with no role for evidence, much less for its critical assessment, in this decision (as nothing is said about it). While waiting for the test result "we find a systematic review of several studies of this diagnostic test . . . [and] decide that it is valid . . ." (p. 72). The results are about the test's "sensitivity" and "specificity" for iron deficiency anemia. The review is an actual one (p. 73), and it is (unjustifiably) taken to be valid on account of affirmative answers to four simple (but inappropriate) questions (propos. IV – 1.10, 11). The test result on the patient is received, and it is (barely) positive in the meaning of positive in that review.

The example shows the calculation of the likelihood ratio (estimate) for the positive result, based on the (empirical values for) "sensitivity" and "specificity" and, then, the calculation of the corresponding post-test probability, given the pre-test probability together with the LR estimate (= 6). The result is 86% (p. 74). (That the result was just barely positive is ignored in this calculation.)

The example is taken up for a second time under Multilevel Likelihood Ratios. There the patient's test result (60 mmol/L) falls in the "neutral" range (35 – 64 mmol/L), for which LR = 1 (as distinct from 6 above), according to a table the source of which (if any) is left unspecified (p. 77). In these terms the test result does not change the diagnostic probability.

Implicitly (but unjustifiably; cf. above), all three of these measures of a test's 'accuracy' are treated as though their values were constant over the different levels of the pre-test probability or, more specifically, of the possible pre-test profiles.

This example fails to illustrate EBM as "the integration of best research evidence with clinical expertise and patient values" (p. 1) – as "critically apprising [the best evidence] for its validity (closeness to the truth), impact (size of the effect), and applicability (usefulness in our clinical practice) . . . and integrating the critical apprisal with our clinical expertise and with our patient's unique biology, values and circumstances [to be followed by] evaluating our effectiveness and efficiency in executing [those steps] and seeking ways to improve them both for the next time" (p. 4).

The closer we look at the EBM teachings, the more clearly "we find," it seems, "a complete system of illusions and fallacies, closely connected with each other and depending on grand general principles . . ." (propos. I – 2.9).

Diagnostic Research: A Paradigmatic Study

A quarter-century ago was published this report:

> Pozen MW, D'Agostino RB, Selker HB, et alii. A predictive instrument to improve coronary-care-unit admission practices in acute ischemic heart disease. A prospective multicenter clinical trial. NEJM 1984; 310: 1273–8.

The report's Abstract includes this:

> In this study ... we sought to develop a diagnostic aid to help emergency room physicians to reduce the number of their CCU [coronary care unit] admissions of patients without acute cardiac ischemia. From data on 2801 patients, we developed a predictive instrument for use in a hand-held programmable calculator, which requires only 20 seconds to compute a patient's probability of having acute cardiac ischemia.

And a part of the rest of the Abstract is this:

> In a prospective trial that included 2320 patients ..., physicians' diagnostic specificity for acute ischemia increased when the probability value determined by the instrument was made available to them. ... Among study patients with a final diagnosis of "not acute ischemia," the number of CCU admissions decreased 30 per cent, without any increase in missed diagnosis of ischemia.

As is evident from this, the avant-garde of diagnostic research understood a quarter-century ago already, that the core mission in this research is to produce empirical *diagnostic probability functions*, DPFs, which can be used to provide critically important probability inputs to decisions (about, e.g., CCU admissions); and that the application of a DPF generally requires *technological development* – in those early days the DPFs' programming into hand-held calculators.

These understandings in this study are so important, and have subsequently been so eminently ignored, that dealing here with this study's particulars – as for its DPF result and of the genesis of this – is not warranted, even though there were problems with these (and also with the trial to evaluate the instrument). For it would be a counterproductive distraction from the paradigmatic essence of that diagnostic study by Pozen et alii. Suffice it to note, first, that (successive) instances of the defined presentation *domain* – based on "chief complaint" and gender-specific range of age – were identified in the emergency rooms of the participating hospitals, and that, in such instances, informed consent for participation in the study was sought. Given the consent, the *data* on an inclusive set of diagnostic indicators, together with the ultimate datum on the presence/absence of cardiac ischemia, were abstracted. A logistic *probability model* (for myocardial ischemia) was fitted to these data – with data-guided reduction from 59 indicators to the final seven. (Cf. propos. III – 2.5–7.)

Diagnostic Research: Paradigm Lost, Example Series I

A series of very eminent diagnostic studies, conducted "under the auspices of the National Heart, Lung, and Blood Institute" of the U.S. and initiated subsequent to the (paradigmatic) study by Pozen et alii, above, started with this one:

> The PIOPED Investigators. Value of ventilation/perfusion scan in acute pulmonary embolism. Results of the Prospective Investigation of Pulmonary Embolism Diagnosis (PIOPED). JAMA 1990; 263: 2753–9.

The synopsis in the report includes this:

> To determine the sensitivities and specificities of ventilation/perfusion [V/Q] lung scans for acute pulmonary embolism [PE], a random sample of 933 of 1493 patients was studied prospectively. .... Almost all patients with [PE] had abnormal scans ... but so did most without [PE] (sensitivity, 98%; specificity, 10%). .... only a minority with [PE] had high-probability scans (sensitivity, 41%; specificity, 97%). ... Clinical assessment combined with [V/Q] scan established the diagnosis or exclusion of [PE] only for a minority of patients ...

The study *domain*, wholly undefined in the synopsis, is orientationally implicit in a point in the Methods section of the report proper, namely that "all patients for whom a request for a V/Q scan or pulmonary angiogram was made were considered for study entry"; and among the eligibility criteria is said to have been that "symptoms that suggested [PE] were present within 24 hours of study entry." There is, however, no specification of these symptoms nor, most notably, of the very importantly domain-defining 'chief complaint' to go with the age criterion (of 18 years or older). There thus was no meaningful definition of the domain of this study (cf. propos. III – 2.11).

With this (ill-defined) domain as their referent, the *objects* of the study evidently had to do with the V/Q scan's "sensitivity" and "specificity" for PE, each of these measures addressed with varying definitions of the scan's positive result, that implying "high probability" (of PE), for example. The scan findings translating into "high probability" are specified, in the Methods section of the report, as also are those for "intermediate," "low," and "very low" probability; but the respective ranges of probability are left without numerical specification.

With any given definition of the range of positive result of the V/Q scan – "high probability" of PE, for example – the object of study was the probability of this positive result, separately for those with and those without PE (as determined by angiography) but with no regard for this probability's dependence on the diagnostic indicators that are available to consider before the V/Q test.

The diagnostic indicators other than the result of the V/Q test, were translated by the "clinical investigators" into "clinical impressions," though "without standardized diagnostic algorithms." Those impressions were expressed in numerically specified ranges of "clinical science probability" (*sic*). On this basis, a supplementary object of study was the prevalence of PE as a joint function of the respective levels of probability based separately on the V/Q test and the "clinical impression" – though not as a formal function but as a matter of empirical proportions specific to the cross-classification categories.

This report drew a response:

> Miettinen OS, Henschke CI, Yankelevitz DF. Evaluation of diagnostic imaging tests: diagnostic probability estimation. J Clin Epidemiol 1998; 51: 1293–8.

This response questioned the rationale for the PIOPED use of those result categories for the V/Q test, constituting a unidimensional ordinal scale and one defined a priori. Its authors suggested that it would be "much more natural to take the development of test-based categories of illness probability ('high probability,' etc.) – insofar as they are of interest at all – to be the first-order objective of the study rather than an *a priori* constraint for it." And they asked, "is it not better to ignore completely such categories [and merely] ask the question of how, in the domain of the study, the prevalence of the illness is a joint function of the readings on the images, possibly together with documented diagnostic indicators other than those from the imaging?"

These authors reanalyzed the 'raw' data from the PIOPED, producing VQ-based diagnostic probability function (DPF) for PE, one that is exclusively evidence-based – in contrast to those a-priori definitions of the findings that jointly imply, for example, unspecified "high probability" of PE. This DPF "placed 29% of the patients in the greater than 60% probability range ... and 60% of them in the less than 20% range ... thus leaving only 11% in the intermediate, 20 to 60% range." By contrast, in terms of "the *a priori* classification used in the PIOPED ... the majority, 62%, of the patients fell in the 'intermediate' probability category, in which the prevalence of PE was 25%; and only 20% of the patients fell in the two lowest-probability categories, representing, approximately, the less than 20% probability range."

These authors also pointed out that "For the purposes of the ultimate diagnosis, the [DPF] that addresses the [pre-test] information jointly with the imaging information (readings) is an obvious extension of what is presented here." The extension was, however, impossible for these authors to implement, as that other information, most remarkably, had not been recorded in the PIOPED.

Then came from the *PIOPED II Investigators* this report (on a study again sponsored by the NHLBI):

Stein PD, Fowler SE, Goodman LR, et alii. Multi-detector computed tomography for acute pulmonary embolism. NEJM 2006; 354: 2317–27.

This was a description of a study that had been prompted by innovations in imaging technology subsequent to the original PIOPED, rather than by any change in the way the investigators had come to think about diagnostic research involving imaging-based inputs into the diagnosis. This report does not dispute what had been said in the critical response (above) to the report on the original PIOPED, including on the basis of reanalysis of its data; all of this is simply ignored.

And this series is continuing further:

Stein PD, Guttschalk A, Sostman HD, et alii. Methods of Prospective Investigation of Pulmonary Embolism Diagnosis III (PIOPED III). Semin Nucl Med 2008; 38: 462–70.

"The purpose of the [PIOPED III] study is to estimate the diagnostic accuracy of [each of two novel imaging tests] for the diagnosis of acute pulmonary embolism (PE). ... The diagnostic accuracy of [one of the tests] alone or the combination of [the two] will be expressed as the sensitivity, specificity, likelihood ratio for a positive test, and likelihood ratio for a negative test."

Now, if the future ones in the presumably ongoing series of PIOPED studies – each of these, too, focusing on the most recent type of imaging for diagnosis about PE – were to conform to 'normal science' (Kuhn) not in the (self-referential) meaning of the original PIOPED as the paradigm but with the study of Pozen et alii (above) in this role, the fundamental new features would be these: The domain of study would be explicitly defined, and not by referral for radiography nor by suspicion of (the presence of) PE but by a particular patient presentation (propos. III – 2.11); and for this domain, each study would address a (carefully designed) pair of diagnostic probability functions (logistic), one involving the diagnostic indicators available before the imaging(s) and the other involving these together with those based on the imaging(s) (propos. III – 2.2). The latter function would serve the decision about the imaging, as it allows determination of the range of possible post-imaging probabilities for the presence of PE, given the pre-imaging profile of the patient (propos. III – 2.3).

Given a research focus on the imaging novelty *du jour* as a source of inputs into diagnosis about PE, as in the PIOPED series of studies, and specifically focus on what bears on informed decisions about undertaking the imaging(s), there is a need to go a bit beyond the Pozen et alii paradigm: The imaging-based terms in the post-imaging function constitute a scoring function for the result of the imaging(s), taking on a particular value in the range of its possible values if the imaging(s) actually is (are) undertaken. Different ranges of realization imply their corresponding ranges of the post-imaging probability of PE being present (per the post-imaging function). Needed for the decision about the imaging(s) thus is information (if not actual knowledge) about the distribution of the score (S), specifically about $Pr(S > s)$ for various values of the cut-off point in this, conditionally on the pre-imaging profile of the patient. The corresponding research need is to study that $Pr(S > s)$ as a function of the pre-imaging indicators, for various values of the range-defining realization (cf. propos. III – 2.4).

All of this reduces to a simple, important point: Research on diagnostic-imaging inputs into diagnosis, exemplified by the PIOPED series of studies, should undergo a *paradigm shift* – leaving behind the focus on the imaging's 'accuracy' in terms of 'sensitivity' and the like, and embracing as the new paradigm the way Pozen et alii (NEJM 1984; 310: 1273–8) integrated the information from electrocardiography with that from history-taking and physical examination in a probability function for diagnosis about myocardial ischemia. Of particular note for the PIOPED investigators – and 'clinical epidemiologists' everywhere – to weigh and consider is this: purported measures of the 'accuracy' of the ECG test in the diagnosis about acute myocardial ischemia had no role in the work of Pozen et alii.

Diagnostic Research: Paradigm Lost, Example Series II

Soon after the introduction of the PIOPED series of studies (above), sponsored by the NHLBI, the NCI (National Cancer Institute) of the U.S. sponsored the Radiology Diagnostic Oncology Group (RDOG) to conduct "comparative studies of the ability of diagnostic imaging modalities to enable the staging of various types of cancer . . ." (ref. 1). The leit motif in RDOG was, akin to that in the PIOPED series, this: "The

clinical [*sic*] evaluation of the accuracy of a diagnostic modality is a key component in the overall assessment of the modality" (ref. 1) − a matter of 'sensitivity,'
'specificity,' and (the area under the) 'receiver operator characteristic.'

In the RDOG research, compared modalities of imaging were applied in parallel
on each subject in the studies (which cannot be done for comparison of effects). The
lead investigators explained that (ref. 1):

> The alternative to the paired design would be a design in which patients are ran-
> domized to one of the imaging examinations being compared. It is difficult to
> imagine that such a design would be feasible in practice; referring physicians do
> not . . . enrol their patients in such protocols. . . . Thus, imaging trials cannot be
> expected to provide information on the long-term consequences resulting from
> the introduction of a new technology. Moreover, because many steps intervene
> between the initial diagnostic evaluation and long-term outcomes, it is diffi-
> cult, if not impossible, to impute the value of specific imaging examinations for
> health outcomes.

This statement of those RDOG authors notwithstanding, the NCI some years
later put out a call for Cooperative Trials in Diagnostic Imaging (ref. 2), in which it
made this assertion:

> More accurate images by themselves will not necessarily motivate new equip-
> ment purchases without evidence that the greater accuracy will translate into
> cost savings or better clinical results. These kinds of endpoints are most
> persuasively assessed using rigorous clinical trials methodology. . . . Where
> appropriate, this evaluation should include estimates of the relative cost-
> effectiveness of diagnostic interventions [*sic*] and of their impact on quality
> of life.

This call led to the formation of the American College of Radiology Imaging
Network (ref. 3). "The specific objectives of ACRIN include . . . [assessment of]
imaging technologies beyond the evaluation of accuracy to include such end points
as the effect of imaging examinations on medical diagnosis, treatment, and health
care outcomes, including quality of life and health care costs."

The erstwhile principals of the ACRIN program recently had a somewhat different tone (ref. 4): "End points beyond accuracy are essential [*sic*] to proving the
value of diagnostic imaging technologies. However, in many situations it will not
be feasible to conduct trials to assess the impact of a diagnostic modality on patient
outcomes."

Now, let us weigh and consider (propos. I − 1.1) the NCI idea (above) that the
"clinical results" of "diagnostic interventions" − results on "quality of life," for
example − "are most persuasively assessed using rigorous clinical trials methodology." It is, for a start, a serious 'category error' to think of diagnostic imaging − or
any diagnostic testing − as an intervention. A diagnostic test is supposed to provide
information (usually about the presence/absence of the illnesses in the differential-
diagnostic set), while a clinical intervention is supposed to change the course of
health for the better. That the application of a *diagnostic test has no effect* on the

course of health would become manifest by means of truly "rigorous clinical trials methodology." In such a trial, the study subjects would be randomly assigned to being subjected to an imaging test or to a sham resemblance of this, with both the patients and the investigators (and everyone else) kept 'blind' to the nature of the assignment and to the test result in each instance. The health effects, if any, of the verum imaging would take manifestation in such a "rigorous" trial; but the result of such a 'thought experiment' already – made arbitrarily large – is resoundingly negative about the test's effect on the course of health.

When the theory of a particular line of diagnostic research is amiss even more profoundly than that underpinning the PIOPED series of studies (and RDOG), examination of particular example studies in it is not warranted.

References
1. Gatsonis C, McNeil RJ. Collaborative evaluations of diagnostic tests: experience of the Radiology Diagnostic Oncology Group. Radiology 1990; 175: 571–5.
2. NIH Guide, vol. 26, Aug 22, 1997.
3. Hillman BJ, Gatsonis C, Sullivan DC. American College of Radiology Imaging Network: new national collaborative group for conducting clinical trials of medical imaging technologies. Radiol 1999; 213: 641–5.
4. Hillman BJ, Gatsonis CA. When is the right time to conduct a trial of a diagnostic imaging technology? Radiol 2008; 248: 12–5.

Etiognostic Research: Clinical Examples

The students in one of the three discussion groups in this course chose for review in class (during the last, fourth week of the full-time course) this report (as one of the total of nine that were thus chosen for review):

Ho PM, Maddox TM, Wang L, et alii. Risk of adverse outcomes associated with concomitant use of clopidogrel and proton pump inhibitors following acute coronary syndrome. JAMA 2009; 301: 937–44.

As it happened, another one of the discussion groups chose an intervention-prognostic study also involving clopidogrel (a platelet aggregation antagonist, commonly known as Plavix); and both of these studies have the pedagogic virtue of representing, arguably at least, the current state-of-the-art in their respective categories of quintessentially 'applied' clinical research, recently peer-approved for publication in one of the preeminent medical journals.

According to the *Abstract* in the report, the Context of the study was this: "Prior mechanistic studies reported that [one of the proton pump inhibitor, PPI, medications used to prevent gastrointestinal bleeds from, e.g., clopidogrel use] decreases the platelet inhibitory effect of clopidogrel, yet the clinical significance of these findings is not clear."

In this Context, the Objective of the study is said to have been, "To assess outcomes of patients taking clopidogrel with or without proton pump inhibitor (PPI) after hospitalization for acute coronary syndrome (ACS) [myocardial infarction or unstable angina]."

The Main Outcome Measures are specified as "All-cause mortality or rehospital-ization for ACS."

The study's "Design, Setting, and Patients" are in the Abstract described this way: "Retrospective cohort study of 8205 patients with ACS taking clopidogrel after discharge from 127 Veterans Affairs hospitals between October 1, 2003, and January 31, 2006. Vital status information was available for all patients through September 30, 2006."

In the Abstract's section on Results, quite extensive, most notable are these two sentences: "In multivariate analyses, use of clopidogrel plus PPI was associated with an increased risk of death or rehospitalization for ACS compared with use of clopidogrel without PPI (adjusted odds ratio [AOR], 1.25; 95% confidence interval [CI], 1.11 – 1.41). ... The association between use of clopidogrel plus PPI and increased risk of adverse outcomes also was consistent using a nested case-control study design (AOR, 1.32; 95% CI, 1.14 – 1.54)."

As Conclusion is said this: "Concomitant use of clopidogrel and PPI after hos-pital discharge for ACS was associated with an increased risk of adverse outcomes than [sic] use of clopidogrel without PPI, suggesting that PPI may be associated with attenuation of benefits of clopidogrel after ACS."

For broadest orientation here, it may well be quite unclear, even to a suitably learned reader, whether this was an etiognostic study or, instead, a prognostic one. The Context statement in the Abstract (above) implies that, in patients treated with clopidogrel, the use of PPI medications can be etiogenetic to such health events as clopidogrel use is intended to prevent (as it reduces the preventive effectiveness of clopidogrel use). The article's title, however, implies a prognostic study, per-haps a merely descriptive – intervention-conditional – one; and the stated Objective in the Abstract also has this prognostic air. The stated methodologic involvements of a "cohort study" and a "nested case-control study," in turn, definitely point to an etiognostic study. On the other hand, again, the routine, in the Abstract (and throughout the article), of writing about "association" instead of effect suggests that, in the minds of the authors, the study was neither etiognostic nor intervention-prognostic. But then again, why all the multivariate adjustments for codeterminants of the events' occurrence if at issue wasn't the events' occurrence in causal relation to their antecedent PPI use? And without intended causal interpretability, study of the "association" (between PPI use and adverse health events) is clinically mean-ingless and unconnected to the Context. All in all, despite all the obfuscation in the writing, at issue must have been *causal* connection between the "adverse outcomes" and their antecedent PPI medication use in the domain of clopidogrel use in sta-tus post ACS (ultimately the "clinical significance" of this; cf. Context above). And a study testing a hypothesis about a causal connection between an adverse health outcome and an antecedent of this generally is a study about etiology/etiogenesis.

Regarding the distinction between an etiognostic study on iatrogenesis (having to do with a medical action) and an intervention-prognostic study, some further clar-ification (as a reminder) may be in order. In an etiognostic study, at issue is *current* (at etiognostic T_0) occurrence (of an adverse event/state) in causal relation to *ret-rospective* divergence in a determinant (etiogenetic; propos. III – 3.18), while in an intervention-prognostic study the object is prospective (as of prognostic T_0) occurrence in causal relation to prospective divergence in a determinant (interventive; propos. III – 4.3). In an etiognostic study the causal determinant's index category can have any time-course whatsoever, while in an intervention-prognostic study its causal/index category is a defined algorithm from prognostic T_0 forward as long as it may bear on the health event/state at issue (propos. III – 4.17). A study that merely tests the hypothesis of a causal connection between an adverse phenomenon of health and an antecedent of this is, by definition, a study on etiology/etiogenesis; and to the extent that a causal rate ratio is quantified, such a study serves (the development of the knowledge-base of) etiognosis and not prognosis.

The study at issue here addressed the occurrence of certain events of health in relation to antecedent PPI use (in the domain of clopidogrel use occasioned by status post ACS); it did not involve defined algorithms of intervention (prospectively as of hospital 'discharge' or any other point in time); and ratios of incidence densities (termed "odds ratios") were quantified. Thus, given that the study was about causality (cf. above), it must be taken to have been an *etiognostic* – rather than intervention-prognostic – study (cf. above).

The true "objective" of the study actually was – as it always is – to document experience of the form of its *object*. In the object of study – occurrence relation (propos. II – 2.12) – the outcomes (not "outcome measures") should have been particular thrombotic cardiovascular events, and possibly also the union of these events. Hospitalization for a health event is not a health event (but an action possibly resulting from it). Nor is the mortality from a health event a health event (but a rate of death from it). And "all-cause mortality" is of no consequence in the context at issue here. In its stead, addressed should have been (the occurrence of any) fatal thrombotic event.

The study's *design* naturally involved specification of its "setting" and "patients" among other features; they were not matters extrinsic to it. Used as the setting was, quite justifiably, a population – cohort – with quite common use (prospective) of clopidogrel and, also, of PPI medication as a supplement to this. Using this cohort, a "cohort study" was designed in its specifics and supplemented by a "nested case-control study," also designed in its specifics. This duality reflects the 'cohort' and 'trohoc' fallacies in epidemiologists' etiologic research (propos. III – 3.19, 23). It deserves note that any properly construed etiologic/etiogenetic study – the study base of this – is 'nested' in a defined source base (propos. III – 3.8).

In the context of the cohort – the source cohort (propos. III – 3.8) – the *source base* should have been understood to be the population-time of this cohort's followup (before the outcome event at issue). With a case series identified from this source base and a sample of it (a series of person-moments from it) drawn, these two series should have been reduced to the corresponding series from the actual *study base* (propos. III – 3.8). This belonging in the study base would have required recent use of clopidogrel together with either index or reference history, as of that moment, of PPI use (propos. III – 3.8). The index history of principal interest would have been PPI use throughout the period (in the past) where PPI use could counteract the effect of clopidogrel use in the prevention of the thrombotic (incl. thromboembolic) event at that time, and the appropriate reference history would have been no use of PPI medications in this period of retrospective time (as of the person-moment at issue). Given these two series, a logistic model should have been fitted to the 'data' on them (propos. III – 3.7). This contrasts with the (perverted, trohoc-type) approach in the "nested case-control study": "Medication use with clopidogrel plus PPI, clopidogrel without PPI, PPI without clopidogrel, or neither of these medications at the time of an event was compared between cases and controls."

Further on the methods, "Based on a sample size of 8205 patients taking clopidogrel after discharge with or without PPI, the minimum detectable odds ratio (OR) with 80% power in a 2-sided test and an α level of .05 (based on an exposure prevalence of approximately 60% and event rate in the nonexposure group of 20%) was 1.7." Now, that 8205 was the size of the source cohort under follow-up as of cohort T_0 and not later, which, together with the unspecified average duration of follow-up leaves the size of the source base (its population-time) unspecified. The "exposure prevalence of about 60%" appears to refer to the fact that "63.9% [of the 8205] were prescribed PPI at discharge, during follow-up, or both," and not to the prevalence of recent PPI use in the source base at large. The "event rate in the nonexposure group of 20%" appears to refer to the fact that "Death or rehospitalization for ACS occurred in 20.8% ... of patients prescribed clopidogrel [at cohort T_0] without PPI [prescription at T_0 or later]," which does not translate to the expected number of outcome events in the source base. This set of statements makes no sense as a specification of "How sample size was determined" (propos. III – 4.27), nor does, by the way, a "2-sided test."

But more importantly, 'sample size determination,' whatever it may have been, or its absence for that matter, in this study or any other, has no bearing on the evidentiary burden of the study's results (propos. III – 4.27). Indeed, the authors of this study report make no point of its relevance in their study; and as this report illustrates, by almost whatever dissemblance authors succeed in satisfying editors' unreasonable requirement of reporting "How sample size was determined" (propos. III – 4.27).

The reported results – whether from "Cox proportional hazard models [with] time-varying covariates and outcomes" in the "cohort study" or from "conditional multivariable logistic regression" in the "nested case-control study" – should be understood not to be about "odds ratio" but about *incidence-density ratio* (propos. III – 3.6, 7). And they are not simply about the degree to which PPI use in the context of clopidogrel use is "associated with" the events at issue (rather than risks for these); they are (by both objects and methods designs) empirical values for *causal* rate (incidence-density) ratios.

There should not have been any *conclusion* (propos. III – 1.1). But if a conclusion nevertheless was to be drawn (per a demand – unreasonable – of the journal), it should not have been about association but about a causal relation, and it should have been expressed in the present tense (as is true of scientific ideas in general) and not in the past tense (which is apposite for statements of mere evidence).

The report's Comment section is largely about "the hypothesis that the interaction of PPI and clopidogrel, rather than PPI itself, was [*sic*] associated with increased adverse outcomes." But, presumably, the *hypothesis* actually was that PPI use is causal (etiologic, etiogenetic) to thrombotic events among users of clopidogrel in status post ACS (as the preventive effectiveness of clopidogrel use is reduced by concomitant PPI use; cf. above). By every reasonable presumption, the hypothesis was not that these two treatments, or medications in them, "interact" – each influencing the other (as in: 'Love makes time pass; time makes love pass'). And it is not that the investigators "found a significant association between treatment with clopidogrel and PPI and the primary combined outcome … " Rather, their study produced *statistically* significant – but otherwise flawed – *evidence* in support of the hypothesis. (Investigators should resolutely stop reporting that they "found" this or that; and even more resolutely, that what they "found" was "significant.")

The Comment section does not address, at all, the causal meaning of the empirical association they reported on. In this study, with a retrospective study base, the risk indicators – potential confounders – were quite superficially documented and, thus, incompletely controlled. (Superficiality in the documentation and control of confounders is commonplace also in epidemiologists' etiologic studies for community medicine, in which PPI use is not a concern.)

As this was a study in the nature of hypothesis testing, quantification of causal rate ratios as functions of their modifiers was not yet a timely concern. It becomes timely if and when the hypothesis – the qualitative idea – becomes more-or-less established. At such a time, suitably specific quantitative knowledge about the thrombogenic effect of PPI use in patients using clopidogrel (as an antithrombogenic medication) needs to be acquired; and this needs to be supplemented with knowledge, again quantitative and suitably specific, about the hemorrhage-preventing effect of PPI use. In all of this research, there is to be orientational clarity on whether being served by it is etiognosis or prognosis (cf. above.)

For research on medicational iatrogenesis of illness – which is clinical research, not epidemiological, its common classification as 'pharmaco-epidemiology' notwithstanding – one of the principal centers now indisputably is in the University of Pennsylvania (the academic home of B. L. Strom, i.a.). An example of the recent work there is this:

Lewis JD, Strom BL, Localio AR, et alii. Moderate and high affinity serotonin uptake inhibitors increase the risk of upper gastrointestinal toxicity. Pharmacoepidemiol Drug Safety 2008; 17: 328–35.

The essence of the report's Summary is this:

Objective … This study examined the effect of medications that inhibit serotonin uptake on upper gastrointestinal toxicity.

Methods … case subjects hospitalized for upper gastrointestinal bleeding, perforation, or benign gastric outlet obstruction were recruited … [and] … control subjects were recruited by random digit dialing from the same region. …

Exposure to medications required use on at least 1 day during the week prior to the index date.

Results ... After adjusting for potential confounders, MHA-SRI use was associated with a significantly increased odds of hospitalization for upper gastrointestinal toxicity (adjusted OR = 2.0, 95% CI 1.4 – 3.0). ...
Conclusions Use of MHA-SRIs is associated with an increased risk of hospitalization for upper gastrointestinal toxicity.

As usual, considerable editing is called for. For a start, the *title* of a report on a scientific study normally – and properly – is not of that declarative form. Instead, it specifies the object of study, though only in very broad, orientational terms (as is done, though without clarity, in the etiognostic study report addressed above). The title should not, as in this case, purport to announce a piece of news about what the study has "found" by way of new knowledge; for, the product of a piece of clinical research is not knowledge but merely evidence bearing on (the advancement of) this (propos. III – 1.1).

The true *objective* of the study actually was, as always in gnostic clinical research, to study – to produce evidence on – the *object* of study. In the latter, it appears from the statement on methods, the outcome at issue was "upper gastrointestinal bleeding, perforation, or benign gastric outflow obstruction," which in truth is an illness composite, not a type of (upper gastrointestinal) "toxicity." But the outcome actually was, per the statement on results, hospitalization for it (just as for the health events in the study addressed above), which is not a health event but an action consequent to it. The stated object of an etiognostic study should specify – even in the broad, orientational terms of the Summary or Abstract – not only the outcome and the etiogenetic determinant in terms of the factor involved (here MHA-SRI use), but also the temporal relation between the outcome and its antecedent period of its studied etiogenesis (by the factor). This is not accomplished by reference, under methods, to the medication's use "on at least 1 day" in a one-week period prior to an unspecified "index date." The domain of the object of study also should be, but in no way is, specified, anywhere in that Summary. The measure of the occurrence relation (causal) in the object of study should be understood to be rate ratio (propos. II – 1.29) – generally incidence-density ratio, IDR (propos. III – 3.6), including here – and not "odds ratio."

The principal *result* actually was not what it is said to have been. An empirical IDR, whose positive deviation from unity, even in a valid etiognostic study, is statistically significant, is but an indication of the corresponding parameter value (causal) also exceeding unity (to some unknown extent); statistical significance of the empirical value for the IDR (its deviation from unity), whatever be its value, does not represent "significantly increased odds of hospitalization for [the outcome]." A suitably edited version of this might have been: After adjusting for the set of potential confounders, the outcome had a statistically significant association with antecedent MHA-SRI use (rate ratio ...). (There may have been no effect behind this association, much less a significant one.)

The Conclusion should have been edited out; but if not, then it perhaps should have been edited to one about the evidence (rather than the abstract truth), such as this: The principal result points to a causal connection, to use of MHA-SRI being causal to the aggregate of gastrointestinal illnesses.

Given this understanding of the report's Summary, the reader's first-order further concern properly is to gain clarity on the *temporal relation* in the object of study, as this is revealed by the full description of the methods of study. Said under Methods in the report proper is this: "For cases, the hospitalization date was the 'index date;' for controls, the interview date was the 'index date.' All exposures were measured backwards from the index date." This the critical reader should find quite problematic.

For one, when a patient is hospitalized for one of the illnesses constituting the composite outcome, scarcely is the period of the potential iatrogenesis – if any – of this outcome confined to the week immediately antecedent to the hospitalization.

A related problem is the nature of the *contrast* in reference to that problematic period of time: Why use "at least 1 a day" versus no use in that week? Why not regular use versus no use in that week? Use of the medication only during the 24 hours just prior to the hospitalization scarcely was causal to the outcome leading to the hospitalization; nor is one-day use a close correlate of use in the entirety of etiognostically relevant time – different from regular use throughout the one-week period.

With the etiogenetic period and the determinant contrast in reference to this suitability designed, the rest of the study's object design should have allowed for confounding (if not for modification of the magnitude of the causal rate-ratio) by *earlier use* of the medication (App. 4: Teacher's response to part D of Assignment 7), along with other confounders (if not potentially substantial modifiers) of the incidence-density ratio.

Actual study of such an object of study inescapably required a series of cases of the outcome event – in principle all of the cases in a defined study base (of population-time) – and a fair sample of that study base; suitable documentation of the case and base series; and fitting the logistic counterpart of the designed model for incidence density to these data – the general structure of the etiogenetic study in the (usual) context of an event-type outcome (propos. III – 3.7–9).

This indeed was the structure of the study in question here; but it is said (in the report) to have been a *case-control* study (presumably as distinct from cohort studies), meaning one in which "case subjects" are compared with "control subjects," first in respect to features that have to do with their (degree of) 'comparability.' This feature of the 'trohoc fallacy' (propos. III – 3.23) indeed was there: a table addresses "Characteristics of the study population" in respect to 35 topics, separately for "cases" and "controls," implying that these two are the constituents of the study population and that 'comparability' (degree of similarity) of their characteristics matters (for validity).

In this comparison of 'cases' with 'controls' the idea generally is that material differences need to be 'adjusted for' in the 'data analysis' (generally by allowance for them in the logistic model). But if this were true, a valid etiogenetic study ultimately based on "case subjects" (a case series) and "control subjects" (base series) would be impossible. For, cases of the outcome occur in high-risk people, so that their occurrence is generally preceded by features associated with high risk – unknown as well as known, undocumentable as well as documentable. The idea is, however, false – a major misunderstanding inherent in the trohoc fallacy that remains commonplace among epidemiologists.

A matter related of 'comparability' of those two series *is* important for the validity of an etiogenetic study, but it is not addressed in that table comparing the "cases" with the "controls": the propensity for errors of documentation of the etiogenetic determinant is to be (essentially) the same for the "case subjects" and "control subjects" (and essentially nonexistent in respect to the potential confounders). Pertaining to this, the table should have documented the distribution of the time lag from the "index week" to the subsequent interview (about the etiogenetic history for that week) for the "case subjects"; and while the lag time for the "control subjects" was uniformly nil, it should not have been. These time-lag distributions should have been similar between the two segments of the "study population" (the two series).

"Cases were obtained from 28 hospitals in Philadelphia and its eight surrounding counties . . . Controls were recruited from the same source population using random digit dialing." "Briefly, to be eligible for inclusion as case or control subject, participants had to reside in the nine-county region, be between 22 and 80 years old, have a telephone, and be able to complete a 30-minute interview."

Now, given that the resident population of the nine-county region seemingly was taken to be the *directly-defined source population* (as distinct from the catchment population of the case-ascertainment process having been the source population, indirectly defined; propos. III – 3.9), the first-stage case series (from the source base) should have been identified comprehensively (though perhaps only in respect to cases severe enough to come to medical attention irrespective of their recent use of medications) from all of the relevant care facilities (for the outcome) for this defined population over a defined period of calendar time; upon each case identification, the patient should have been targeted for interview; and upon targeting, each patient should actually have been interviewed. The report gives no indication of the extent to which this was done.

Correspondingly, a fair sample of the person-moments of that source base should have been drawn – by the use of population registries – and the persons involved in these then actually interviewed in the same period of calendar time as those in the case series (with a view to similar lag-time distributions; cf. above). This evidently wasn't done.

In the reduction of these first-stage series to the corresponding second-stage series, the scientific criteria of the person-moment belonging in the study base (at the time) should have included either the index or reference history (involved in the appropriately designed object of study; cf. above) together with membership in the domain of the object of study (which wasn't specified). And among the 'technical' criteria should have been ones providing assurance of ability to recall medication use (incl. among these, further restriction of the upper bound of the range of age; cf. propos. III – 3.8.)

Among the validity issues of identification, targeting, and actually interviewing, the report in its Discussion section addresses the interviewing, saying that "There is little reason to believe that controls who used antidepressants would be less likely to participate than case subjects who used antidepressants. Likewise, if differential participation were to have occurred, we would have expected to see a similar association with bleeding among the low-affinity non-SSRI medications."

Now, insofar as the study data are used to affirm the study's validity, this needs to be done with the necessary degree of 'mental discipline' (Kant). For the rate ratio and its 95% confidence interval the values for the high-affinity antidepressants (relative to none) were 2.1 and 1.3 – 3.3, while the low-affinity counterparts of these were 1.0 and 0.4 – 2.3. The difference of the logarithms of these ratios is 0.74 with a 95% interval from – 0.3 to 1.7, statistically well consistent with log-difference 0.0 and, thus, with no difference between the high- and low-affinity types of antidepressant.

Regarding the alternative for causality as an explanation for the empirical association, presuming (with considerable reservations; cf. above) that it is descriptively valid in reference to the study base, confounding was controlled by entering the potential confounders (dubious as to the accuracy of their documentations) jointly in the logistic model, as is appropriate. But then, "Variables were selected for the final model if inclusion altered the [rate ratio] for MHA-SRIs by 10% or more." The set retained for control consisted of "age, sex, race, alcohol consumption, history of ulcer disease, and hypertension," these and nothing but these. But there should not have been any data-guided reduction of the set of potential confounders; for in the stepwise reduction, confounding (of the result) is stepwise reintroduced.

The authors considered confounding by indication for the medications' use (i.e., depression) in respect to the result for high-affinity MHA-SRI use, but said: "That we did not observe a similar association with the low-affinity, non SRI antidepressants argues strongly [sic] against confounding by indication as a source of bias." That they "did not observe" a particular association does not mean that it doesn't exist (in the abstract), especially as there was no statistically significant difference between the rates with the two types of antidepressant (cf. above). It really is good to bear in mind the maxim that 'Absence of evidence is not evidence of absence.'

As usual, the causal rate-ratio (incidence-density ratio) was not addressed as a joint function of its modifiers as well as particulars of the index history (propos. II – 2.2, 22); and so, the evidence presented – were it to be viewed as valid for what it did address – would not provide for quantification of the etiognostically relevant risk ratio (causal) for a particular history of antidepressant use and a given subdomain of the study object's domain and, thus, "for the individual patient."

And even if the object had been of that distinction-making type, it would not have been appropriate for a prognostic study (propos. II – 2.27). But the report ends with a remark (quite a gratuitous one) pertaining not to etiognosis but to prognosis: "Whether this risk of bleeding [sic] with MHA-SRIs is sufficiently large enough [sic] to warrant selection of alternative therapies depends on the benefit to risk ratio for the individual patient." The predicate for this is that, "Our study and others document that the relative risk [*sic*] of GI hemorrhage with moderate and high affinity SRIs is elevated."

This report is particularly instructive because it comes from such an eminent source and has to do with what epidemiologic (community-medicine) researchers know best – etiologic/etiogenetic research, that is. There is much for even leading research-epidemiologists yet to learn about this line of genuinely epidemiological research. And accordingly, a certain measure of reserve and even humility would be in order when setting out to teach 'clinical epidemiology' to clinicians (cf. propos. III – 3.24), in respect to research on pharmaco-etiogenesis for clinical etiognosis, for example.

Prognostic Research: Clinical Examples

An eminent example of recent studies bearing on *descriptive* prognosis is this one:

> Defrano R, Guerci AD, Carr JJ, et alii. Coronary calcium as a predictor of coronary events in four racial or ethnic groups. NEJM 2008; 358: 1336–45.

In the report's Abstract the essentials are these:

> *Background.* In white populations, computed tomographic measurements of coronary-artery calcium [CAC] predict coronary heart disease [CHD] independently of traditional coronary risk factors. However, it is not known whether [CAC] predicts [CHD] in other racial or ethnic groups.

> *Methods.* We collected data ... in ... men and women ... [who] had no clinical cardiovascular disease at entry and were followed for a median of 3.8 years.

> *Results.* ... In comparison with participants with no coronary calcium, the adjusted risk of a coronary event was increased by a factor of 7.73 among participants with coronary calcium scores between 101 and 300 and ... ($P < 0.001$ ...). Among the four racial and ethnic groups, a doubling of the calcium score increased the risk of ... any coronary event by 18 to 39%. The areas under the receiver-operator-characteristic [ROC] curves ... were higher when the calcium score was added to the standard risk factors.

> *Conclusions.* The [CAC] score is a strong predictor of incident [CHD] and provides predictive information beyond that provided by standard risk factors in four major racial and ethnic groups in the United States. No major differences among racial and ethnic groups in the predictive value of [CAC] scores were detected.

According to the Background section of the Abstract, implicitly, this study was to address the level of CAC as a *prognostic indicator* regarding CHD, and specifically in domains of non-white persons. In the quotes above these domains are said to be "racial or [*sic*] ethnic" (title and Background) and then "racial and [*sic*] ethnic" (Results and Conclusions) – while simply "ethnic" in the full report's section on Methods, Results, and Discussion. From what is said in the Abstract's section on Methods, however, one can surmise – correctly – that the domain-defining indicator at issue actually was demographic, with the categories "white," "black," "Hispanic," and "Chinese." So, the study was to address a CAC score as an "independent" prognostic indicator about CHD among blacks, Hispanics, and Chinese, the question of whether the CAC score has *marginal informativeness* in these demographic domains (just as it has among whites).

To that question, it seems, the answer should have been affirmative on a-priori grounds, and a more meaningful question would have been whether a demographic indicator has marginal informativeness supplementary to the "traditional" ones augmented by the CAC score. Pertaining to that CAC-related question about the demographic distinctions, the only point in the Abstract's section on Results is the one quoted above, namely that doubling of the CAC score "increased the risk ... by 18 to 39%" among the four demographic subcohorts. The CAC score, however, is but an indicator of the risk (of CHD), and not something that increases (or decreases; i.e., influences) the risk (causally). And regardless, those two numbers do not refer to risk (theoretical) but to something solely empirical: from the body of a table in the report one can learn that some "hazard ratio" was 1.18 for Hispanics and 1.39 for blacks (i.e., 18% and 39% in excess of 1.00, respectively). The meaning of this, in turn, is given in a footnote to that table: "Hazard ratios were calculated with the use of Cox regression for [CHD] ... for baseline levels of $\log_2 (CAC + 1)$ after adjustment for risk factors and interactions between racial or ethnic group and diabetes (only significant interaction). Hazard ratios are [*sic*] calculated on the basis of a doubling of CAC + 1."

Given all of this behind the only numbers in the Results section of the Abstract that bear on the core object of this study, it may merit a passing note that in the Index of what arguably is the leading textbook on EBM (ref. in propos. IV – 1.2) there is no entry for "hazard ratio" or "Cox regression" or "adjustment" or "interaction." The same is true, also, of the corresponding textbook on 'clinical epidemiology' (ref. in sect. V – 4).

In the result of that Cox regression analysis, according to the Results section of the full report, "There was no [evidence of] interaction between ethnic group and the risk associated with increasing CAC score." Presumably meant by this is that, in the regression result, each product term involving a demographic variate and $\log_2 (CAC + 1)$ lacked statistical significance (at the level of $\alpha = 0.10$); that is, that there was, in this meaning, no statistically significant evidence of the marginal informativeness of the CAC score being dependent on the demographic indicator and vice versa. This presumably was, in part at least, the basis for the reported result (*sic*) in the Abstract's section on Conclusions (*sic*) that "No major differences among racial and ethnic groups in the predictive value of [CAC] scores were detected." Absence of evidence, however, is not evidence of absence; and relevant evidence about the absence of that prognostic interdependence would have been interval estimates of the coefficients in those product terms – narrowness of the ranges in these. These intervals were not reported, however.

Having thus concluded (*sic*) that the CAC score, when added to the "traditional" indicators of CHD risk, has marginal informativeness that does not depend on those demographic distinctions, the authors turned to quantification of that informativeness; and in the Abstract's section on Results they report on this first (cf. above). That statement, too, requires editing. "No coronary calcification" is a misrepresentation of CAC score equal to zero; it is not that "adjusted risk" was "increased" to a particular extent but, instead, that the "hazard" (incidence-density) ratio conditional on the other indicators of risk had a particular empirical value; that ratio – notably for the wide range of the CAC score – is not reasonably reported as having had the value "7.73" – as though even the second decimal had some meaning in this context; and insofar as that empirical ratio – 'point estimate' – is taken to be of quantitative interest, as a supplement to it should have been given – a measure of its (im)precision – the width of a corresponding 'interval estimate.'

Having taken this quantitative point of view, and even concluded (*sic*) that the CAC score is, marginally, a "strong predictor" of CHD in all four of the demographic categories, it is utterly meaningless to report – in the Abstract, no less – about the qualitative manifestation of this in ROC curves.

That the study is said to have deployed "a population-based sample" (Abstract) of humans is – presumably as a basis for claiming 'generalizability' or 'external validity' (propos. III – 4.27) – totally meaningless, whatever may be the exact procedural meaning of that term. (In the laboratory, no one claims to have conducted medical research by the use of a "population-based sample" of *rattus norwegicus*, say.)

Now, let us *rethink* this study from the very beginning, from this point of departure: in prognosis about CHD, the indicators might include not only the "traditional" ones (age, gender, . . .) but also a/the CAC score and a certain demographic indicator other than age or gender besides. And let us take it that, in the study, the initial concern is the only in-essence 'applied' one (propos. I – 2.5) to test whether all of these indicators have informativeness about CHD risk conditionally on all of the others being accounted for – marginal informativeness in this meaning.

Let us take the demographic indicator up as an example. Regarding this DI, that initial concern is not to test whether its prognostic informativeness, however quantified, varies according to the CAC score or any other one of its codeterminants of the risk of CHD. The concern is, simply, to test – to produce evidence pertaining to the question – *whether it bears information* about the risk of CHD *conditionally* on the defined codeterminants, any nonzero amount of information.

In this testing, let us adhere to the principle – an adaptation of Occam's razor – that all unnecessary complexity in the form and production of the evidence is to be avoided (to maximize the intelligibility of the evidence to its recipients in the relevant scientific community). In other words, let us heed the principle of keeping the object(s) and methods of the study as *simple* as possible – obviating, if at all possible, the need for statements such as, "Tests for nonproportional hazards [in that Cox regression] using Schoenfeld residuals resulted in nonsignificant findings in all analyses."

So, we take the overall object of study to be the association between the DI at baseline and subsequent CHD event (suitably defined) conditionally on all of those other indicators. And to this end, the need is for stratification of the data in such a way that within each of the strata the subjects in the different categories of the DI have similar – 'comparable' – distributions by all of those codeterminants of the risk. Given such a stratification of the data, the interest is in the intrastratum association between the DI and subsequent CHD, in evidence about this summarized across the strata.

The stratification can be based on a risk score for the CHD event, involving all of those indicators (the DI included) but evaluated at a single category of the DI (white race, say). The needed scoring function is a discriminant between the occurrence and non-occurrence of the CHD event, now most naturally based on (the linear compound from) logistic regression. The recipient of the evidence need not understand this scoring as the basis of the stratification, given that the report includes stratum-specific data on the codeterminants – showing their balanced distributions across the categories of the DI (ref. 1). For any given DI contrast, then, the test of statistical significance of the intrastratum association and the 'point estimate' of the corresponding rate ratio (or odds ratio, if preferred) can be the very familiar ones (ref. 2), with a simple 'interval estimate' based on these (refs. 3, 4).

The relevance of the DI for prognosis about CHD in the absence of the CAC score could, of course, be studied in this same, simple way, so long as the concern is merely to learn whether it deserves to be included as a prognostic indicator jointly with the "traditional" ones. And whatever is true about studying the DI in the presence or absence of the CAC score is just as true, *mutatis mutandis*, about studying the CAC score in the presence or absence of the DI, as for the qualitative question about its (marginal) relevance in quantifying the risk for CHD, about there being any relevance at all. In these studies, the focus can be on whatever comparative measure of CHD occurrence, ratio of proportion-type incidence over an arbitrary (unspecified and varying) span of prospective time, for example (as in the approach outlined above).

Research of this type serves *object design* in subsequent research for quantification of the risk of CHD, in prognosis about CHD. For it obviously bears on what descriptors of the instances from the domain of prognosis to include in the set of prognostic indicators, in the example here in the particular context of the CAC scoring being feasible to do or readily available for incorporation in the profile. The DI, such as it is here, obviously is always feasible to actualize and deploy.

By contrast, though, research on *interrelations* such as whether the DI has bearing on the existence of the prognostic relevance of the CAC index, or on the magnitude of the latter in terms of the ratio of CHD risk, in the context of the "traditional" risk indicators having been accounted for (as in the study by Defrano et alii, above), is *irrelevant* for the design of the PPFs (prognostic probability functions) for future research – to say nothing about relevance to prognostic practice before such research. And again, this is the case, *mutatis mutandis*, the other way around – as for the potential bearing of the CAC index on the relevance of the DI. For, absence of the relevance of a potential indicator (conditionally on another one) is not feasible to demonstrate or even to produce supporting evidence for.

Once there is confidence (in the relevant scientific community) that, apart from the "traditional" indicators of risk, both the DI and an index of CAC belong in a PPF for CHD (for use in settings in which a/the CAC index is accessible), the need is to design the form of that PPF in its details, including as for the interdependencies of (marginal) informativeness among the indicators – and then to suitably study that PPF (à la sect. III – 4).

References
1. Miettinen OS. Stratification by a multivariate confounder score. Am J Epidemiol 1976; 104: 609–20.
2. Mantel N, Haenszel W. Statistical aspects of the analysis of data from retrospective studies of disease. J Nat Cancer Inst 1959; 22: 719–48.
3. Miettinen OS. Simple interval estimation of risk ratio. Am J Epidemiol 1974; 100: 515–6.
4. Miettinen OS. Estimability and estimation in case-referent studies. Am J Epidemiol 1976; 103: 30–6.

One of the three discussion groups in the course chose for review this report (along with two other reports):

The ACTIVE Investigators. Effect of clopidogrel added to aspirin in patients with atrial fibrillation. NEJM 2009; 360: 2066–78.

As at issue is a very fresh report on a very major trial in a preeminent medical journal – and with central involvement of the virtual headquarters of 'clinical epidemiology' and EBM (McMaster University) – the students seem to have wanted to have reviewed a particularly notable example of contemporary realities in prognostic – specifically *intervention*-prognostic – research. So, it here deserves almost a page-by-page review – critical review, that is (as in EBM).

In the Background section of the report's Abstract the authors say this: "We investigated the hypothesis that the addition of clopidogrel [an oral antiplatelet medication, rather like aspirin, commonly known as Plavix] to aspirin would reduce the risk of vascular events in patients with atrial fibrillation." But, the investigation itself is not background to the investigation; and investigated was not a hypothesis (a matter of psychology) but an effect (a matter of biology) – a hypothesized effect.

The Methods section of the Abstract, while specifying the total number of patients enrolled into the trial, leaves it unstated how many were assigned to the clopidogrel and placebo groups, respectively; and while it specifies the dosage of clopidogrel, that of aspirin it leaves unspecified.

Under the Abstract's Results section, the first and main point is that the rate of "major vascular events," viewed as a composite, was lower in the clopidogrel group. The empirical rate-ratio the authors term "relative risk," even though at issue is not a pair of risks (inherently only theoretical) but the empirical counterparts of these. "The difference was primarily due to a reduction in the rate of stroke with clopidogrel," the authors say. This statement about "reduction" is, however, one of inference rather than result, while the corresponding result statement would be about the empirical difference without regard for whether it represents reduction (due to the treatment).

A larger problem with the Results section is this: It opens with a declaration of 'good news' in respect to "major vascular events" but closes with (unsurprising) 'bad news' in respect to "major bleeding." The latter also is a "major vascular event," but it was not included in the composite of this. Had major bleeding been included among major vascular events, as it should have been, the empirical rate of major vascular events among the clopidogrel users evidently would not have been appreciably lower than that in the placebo arm of the trial.

The Abstract should not have the Conclusions section (propos. III – 1.1). But if one nevertheless does proceed to conclude something about the object of study, it should not be expressed in the past tense (appropriate for statements about experience/evidence) but in the present tense (apposite for statements about the timeless/universal/abstract); and its stated referent should be an abstract category – the study domain – and not the study base (particularistic) representing this. At variance with this, the authors write that "the addition of clopidogrel to aspirin [in the trial] reduced [*sic*] the risk of major vascular events ... and increased [*sic*] the risk of major hemorrhage [in the study experience]."

In the introductory section of the article proper (cf. Background above) the authors say that "Adjusted-dose vitamin K antagonists and antiplatelet agents reduce the risk of stroke by 64% and 22%, respectively [ref.]." In reality, however, the degree of risk reduction (in ratio terms) is prone to depend on the prognostic indicators, the particulars of the treatments, and prognostic time; and whatever might be these specifics, the corresponding risk reductions are not knowable with anything like the degree of accuracy implied by those 64% and 22%. Proportion-type empirical rates are being confused with risks (inherently theoretical).

Similarly, it cannot be that "The benefit of combining clopidogrel with aspirin has been proven in patients with acute coronary syndromes [ref.]." Proofs have a place (a central one) in theoretical sciences; but they have no place in empirical sciences, in which the knowledge is, inescapably, uncertain (propos. II – 2.11).

The study object's domain needs to be inferred from the Study Participants subsection under Methods. The admissibility statements there are quite nonspecific, as exemplified by this: "Patients were excluded if they required vitamin K antagonist or clopidogrel or ..." At issue in this are opinions about the treatment of choice, not specified facts pertaining to the prognostic profiles of the patients.

As for the outcomes (of treatment) in the objects of study, the editorial imperative to distinguish between "primary" and "secondary" outcomes (propos. III – 4.27) was heeded: "The primary study outcome was any major vascular event (stroke, myocardial infarction, or death from vascular causes). The most important secondary outcome was stroke. ..." These specifications are, however, strikingly illogical (cf. above). But as they actually have no bearing on the burden of the presented evidence (propos. III – 4.27), the reader is free to proceed from the basic facts that the purpose of adding the clopidogrel element to the use of aspirin in the treatment of atrial fibrillation is to achieve the *intended* effect of reducing the risk of thromboembolic and plainly thrombotic events; that the primary unintended –

adverse – effect of this added element in the treatment is increase in the risk of bleedings; and that what matters in the end is the balance of these countervailing effects in terms of the risk of a "major vascular event" of either type (cf. above).

From this vantage, the reader would reasonably wish to distinguish between the risk of the most major vascular events, namely *fatal* ones – whether thromboembolic, plainly thrombotic, or hemorrhagic – and the aggregate of major *nonfatal* vascular events – the respective effects on these.

In this vein, then, the reader would reasonably take the view that the intended effects need to be thought of as being in their magnitudes substantially dependent on the level of the 'background' risk: in high-risk people the intended effect – in terms of risk difference (propos. III – 4.4) – should be expected to be relatively large, and close to nil in those with only occasional and rather brief episodes of atrial fibrillation and with also otherwise a profile of low risk for thromboembolic or other thrombotic events. The unintended, hemorrhagic effects, by contrast, might not be correspondingly larger in persons at high risk for thrombosis. Thus the overall effect could be expected to be positive/favorable in high-risk people, and negative/unfavorable in low-risk people.

In these terms, the *object* of study for both the fatal and nonfatal types of composite outcome would be the event's incidence-density as a joint function of the prognostic indicators, type of intervention, and prognostic time, to be transformed into the corresponding function for cumulative incidence or risk (propos. II – 2.27). The models should allow for exploring the effect's dependence on the 'background' risk (as outlined above).

A separate model for the adverse effect (on the risk of bleeding) should allow for exploring whether this effect tends to be concentrated in the earliest part of prognostic time (whereupon the susceptibles might be depleted from among those still being treated). Alternatively, or in addition, the models for fatal and nonfatal vascular events should allow for improvement in the net benefit after the earliest part of the prognostic time.

In the methods of the study, an eminent feature was, of course, the patients' random assignment (in equal numbers) of the study subjects to the verum and placebo arms of the trial, with 'double blinding' of the assignment. But, nothing is said in the report about efforts to ensure *adherence* to these assignments, whether in subject selection or after the assignments; and as a matter of Results (*sic*), quite poor rates of adherence are reported.

The editorial imperative of reporting "How sample size was determined" – irrelevant though it is (propos. III – 4.27) – was heeded. Specifically, in the subsection Statistical Analysis (*sic*) under Methods the report says this: "On the basis of an expected annual primary-event rate of 8% among patients treated with aspirin alone, we estimated that enrollment of 7500 patients during a period of 2 years would provide a statistical power of 88% to detect a relative reduction of 15% in the risk of major vascular events with the addition of clopidogrel to aspirin. The study was designed to accrue at least 1600 primary events." This, however, is a statement – incomplete and also otherwise deficient – of what the 'power' – probability of a statistically significant difference – actually was, and not a statement of how 'sample size' was "determined." But no matter: the need was to satisfy editors' unreasonable requirements; and whatever works is good enough for this purpose, though not as a matter of clinical scholarship.

In that (irrelevant-to-report) calculation, it was wholly irrelevant that the subject accrual would take two years; but relevant was the expected degree of adherence to the assigned treatments. Nothing is said about the latter in the report. When considered was a "reduction of 15% in the risk ... with the addition of clopidogrel to aspirin," it was necessary to consider the expected counterpart of this in the context of the expected degree of non-adherence to the assigned treatments.

Next, if corresponding to that unspecified amount of (average) risk reduction there purportedly was to be the probability of 88% to "detect [it]," the meaning actually must be that with this probability the null P-value was going to be less than the test's level of statistical significance (α). But, this level is left unspecified. And it deserves note that $P < \alpha$ would not at all mean that 15% reduction (under full adherence) has been detected. It would mean, only, that on the level α of statistical significance there is evidence (statistical) of *some* difference in the abstract (on the premise of valid genesis of the result; propos. III – 1.5). But this was well known for both the intended and unintended effects before the trial already, and the need merely was for *quantitative* evidence (suitably specific, about the magnitudes of the effects). And to this end, the proper concern in the trial's design – again irrelevant to report on – would have been the results' expected levels of precision (instead of the probability of getting $P < \alpha$).

In the main, as for this topic, it needs to be noted that nowhere in the report is any reference being made to this "sample size determination" (with its arbitrary and incompletely specified inputs), notably as to how it might bear on the evidentiary burden of the study in the contexts of its actual "sample size." So the authors, too, treat it as irrelevant to report on (cf. propos. III – 4.27). And those who are scholarly enough to accord terminology its due regard will note that enrollment of patients into a clinical trial is not sampling and that, therefore, the size of the trial cohort is not the trial's "sample size" (cf. propos. III – 4.27).

As for the *ethics* of the study methods, there was an "independent data and safety monitoring board" with a particular set of stopping rules to potentially implement, and "All patients provided written informed consent before participating in the study." But, potential and actual participants evidently were not informed about the already-accrued evidence from the study itself, in line with the prevailing general culture of pseudo-informed consent for entry into and continued participation in intervention experiments. Had the participants been so informed, there would not have been any need for stopping rules with a view to their safety (as they are the proper decision-makers not only in routine practice but in its experimental counterparts as well; cf. propos. II – 3.6).

In the Results section, no point is made, in the text, about the fact that there was no difference at all in the rates of the most important outcome event, namely "death from vascular causes." The rate ratio ("relative risk") was 1.00 with an associated 95% 'confidence interval' of 0.89 – 1.12 (Table 2). The rate ratio presumably was lower than this in high-risk persons and higher than this in low-risk persons (cf. above), but no data on this are given.

While this (important but unheralded) result is a clear line item in the table with the caption "Relative risks of primary and secondary outcomes, according to treatment group," this table as a whole is quite puzzling. Focusing on the clopidogrel plus aspirin column, the number for any stroke is given as 296, but the sum of the numbers for ischemic, hemorrhagic, and "of uncertain type" strokes is 306 (> 296), and the sum of the numbers for "nondisabling" and "disabling or fatal" strokes is 305 ($\neq$ 306; > 296). And much more importantly, while the number for "death from vascular causes" is 600 and that for nonfatal strokes is 296 – 70, these two adding up to 826, the implication is that only six nonfatal events (832 – 826) were among the 90 cases of myocardial infarction together with 54 cases of "non-central nervous system systemic embolism." The aspirin column is just as puzzling.

This table in conjunction with that for "Relative risks of hemorrhage, according to treatment group" do not allow for identification of the treatment-specific numbers of nonfatal vascular events in the sense of those involved in the "primary outcome" supplemented by those of "major hemorrhage." For the fatal and nonfatal events combined, the last sentence in the Hemorrhage subsection under Results gives the numbers 968 and 996, which imply for the nonfatal events the numbers 968 – 600 = 368 and 996 – 599 = 397 for the verum and placebo arms, respectively. These imply a rate ratio of 0.93 with a 95% (im)precision interval of 0.81 – 1.06.

Thus, in the study's results there isn't any statistically significant (numerical) evidence of clopidogrel having a favorable effect on the risk of even nonfatal "major vascular events" (hemorrhages included), let alone on the risk of fatal ones.

This said, it must be deemed seriously misleading to report, as the main result in the Abstract, a rate ratio of 0.89 and its associated 95% interval of 0.81 – 0.98, with P = 0.0001, for "major vascular events" – on the basis of excluding major nonfatal hemorrhages from among these while apparently including even minor cases of the other types of nonfatal outcome.

In the Subgroup Analyses subsection under Results, a large number of possible "interactions" – this misnomer referring to clopidogrel's effect on the risk of the "primary" and "secondary" outcome varying according to particular indicators of risk – are explored. That the risk indicators are considered one at a time, instead of considering levels of risk defined in multivariate terms, makes this elaborate presentation of generally negative results quite meaningless.

The Discussion section of the report gives no insight into the evidence beyond what is presented under Methods and Results. Instead, it abounds with unjustifiable statements such as: "The addition of clopidogrel to aspirin reduced the rate of major vascular events from 7.6% per year to 6.8%"; and, "In ACTIVE A, clopidogrel plus aspirin reduced the risk of major extracranial hemorrhage by 51% and major intracranial hemorrhage by 87%."

Given a patient with a particular history and status in respect to atrial fibrillation and a particular profile in respect to other indicators of risk for thromboembolism, as well as in respect to thrombosis per se and the risk of hemorrhage, and commitment

to treatment with aspirin, for the decision about supplementary treatment with clopidogrel the need is to know about the magnitudes of these profile-specific risks as
functions of prognostic time and how they depend on the use/non-use of clopidogrel. The data from this study could be used to produce such risk-estimate functions,
as recently described (propos. III – 4.12–13).

Another one of the three discussion groups in the course chose for review this
report:

> POISE Study Group. Effects of extended-release metoprolol succinate in
> patients undergoing non-cardiac surgery (POISE trial): a randomized controlled
> trial. Lancet 2008; 371: 1839–47.

As in the case of the study addressed above, at issue here also is a fresh report on
a major trial in a pre-eminent medical journal – and again with central involvement
of the university of the leaders of both 'clinical epidemiology' and the EBM
movement.

In the report's up-front Summary, the *Background* statement is this: "Trials of
β blockers in patients undergoing non-cardiac surgery have reported conflicting
results. This randomized controlled trial, done in 190 hospitals in 23 countries,
was designed to investigate the effects of perioperative β blockers." However, given
those "conflicting results," relevant further background would have been an idea of
why the results have been conflicting and how the conflicts could be resolved. That
which under Background actually is said about this POISE trial would properly
belong under the Methods section of the Summary.

From what is said in the Methods section of the Summary one can infer that the
object of study was the occurrence of "a composite of cardiovascular death, nonfatal myocardial infarction, and non-fatal cardiac arrest" – specified as "the primary
endpoint" – in causal relation to treatment by either the medication of interest or
placebo, "started 2 – 4 h before surgery and continued for 30 days." One is left
wondering why the supplementation to the "cardiovascular death," insofar as there
was to be one in the "primary endpoint," was not *serious* but nonfatal cardiovascular
event, very notably inclusive of stroke of this type.

The presumably principal one of the *Findings* reported in the Summary is the
first one: "Fewer patients in the metoprolol group [n = 4174] than in the placebo
group [n = 4177] reached [*sic*] the primary endpoint (244 [5.8% of the] patients
in the metropolol group *vs* 290 [6.9% of the] patients in the placebo group; hazard
ratio 0.84, 95% CI 0.70 – 0.99; p = 0.399)." The reported other "findings" include
the statistics concerning "more deaths in the metoprolol group" and more strokes
in it also, but the former without any indication of whether those deaths were of
cardiovascular causes and the latter equally puzzling as to whether the strokes at
issue were fatal or non-fatal or either.

· The ensuing *Interpretation* is this: "Our results highlight the risk in assuming a
perioperative β blocker regimen has benefit without substantial harm, and the importance and need for large randomized trials in the perioperative setting. Patients are
unlikely to accept the risks associated with perioperative extended-release metoprolol." So, no "interpretation" is given of the result on "the primary outcome" in

respect to the Background of "conflicting results." Instead, presented are "interpretations" that have nothing to do with this, nor with the evidence from the study. Properly construed, at issue is inference (inductive) on the basis of the evidence from a study – the use of the evidence in updating belief about the object; and, contrary to what journal editors expect, this is not a function for the investigators but of members of the relevant scientific community (propos. III – 4.27).

Different from the Background statement in the Summary, the Introduction to the full report makes no allusion to "conflicting results of prior randomized trials." Instead, "A meta-analysis [of them] suggested that β blockers might prevent major cardiovascular events but increase the risk of hypotension and bradycardia."

In the Statistical Analysis (*sic*) subsection under Methods is this: "Assuming an event rate in the control group of 6% for our primary outcome, we calculated that . . . 10000 patients [would provide] 92% power . . . to detect a relative risk reduction of 25% (two-sided $\alpha = 0.05$) [ref.]." Said actually should have been that with 5000 patients in each of the trial's two arms, the calculated probability of a two-sided (*sic*) test giving $P < 0.05$ was 92% on the premise that the risk ratio is 0.75 – and not that 25% reduction of the risk was going to be detected with 92% probability.

Presented by this is not "how sample size was determined," as is expected by journal editors (propos. III – 4.27), but the 'power' implication of the 'sample size' that was adopted a priori. But no matter: the authors make nothing of this calculation as for the burden of the evidence regarding "the primary outcome," irrelevant to this as it is (*ibidem*).

The rest of the Statistical Analysis subsection under Methods reflects to quite an exceptional extent commitment to frequentist doctrines concerning topics such as "prespecified primary subgroup analysis," "prespecified secondary subgroup analyses," scheduling of interim analyses with their respective "thresholds" for stopping the trial, and how "The α-level for the final analyses remained $\alpha = 0.05$ in view of the infrequent interim analyses, their extremely low α levels, and their requirement for confirmation with subsequent analyses." (Cf. propos. III – 4.27.) Quite a passage for practitioners of EBM to critically evaluate.

Under Results the first point is the remarkable one that "central data consistency checks" and "on-site auditings" indicated that "fraudulent activity had occurred" in respect to $752 + 195 = 947$ of the 9298 participants that were randomized. The data on these participants were excluded from further consideration; but no true assurance is given that the remaining data were not tainted.

The principal result is, in this section, given by this expression: "Significantly fewer participants in the metoprolol group than in the placebo group experienced the primary endpoint (hazard ratio 0.84, 95% CI 0.70 – 0.99, p = 0.00399; . . .). This beneficial effect . . ." So, the result was not that a 25% reduction in the risk had (or hadn't) been detected (cf. above). The "significantly fewer" expression is a misleading way of referring to an empirical difference that merely is, in its deviation. from the null value, statistically significant and, thus, indicative of *some* reduction in the event rate. (For that statistical significance, the null P-value is not justifiably reported with three 'significant' digits.) And even more importantly, this indeed was but an empirical difference, not a "beneficial effect."

The Results section in the report proper, even, leaves unspecified whether the statistically significant difference in the frequency of "deaths" noted in the Summary referred to deaths from cardiovascular causes or from any cause. The relevant table, however, gives this result for "total mortality." As for strokes, the result presented in the Summary without specification of whether they were fatal, nonfatal, or either is equally unspecified in the text in the Results section of the report proper; but in the relevant table that result corresponds to "stroke" rather than "non-fatal stroke." The reader should not have to consult tables to find the meanings of the terms in the text of a report on a piece of research.

A figure "shows the results of our prespecified subgroup analyses and indicates consistency of effects." But there was no "prespecified" model to address the risks as a joint function of risk indicators and intervention, with allowance for the treatment effect's dependence on the value of the multivariate measure of the 'background' risk. This is what the study should have been all about, separately for well-conceptualized composites of fatal and serious-but-nonfatal outcomes.

The report's Discussion section addresses the genesis of the results (re validity) only in respect to "the exclusion of a number of randomized patients from our analyses because of fraudulent activities," which "could be seen as a limitation"; but it asserts, without any particulars, that "our on-site monitoring . . . showed that the trial was rigorously done in all these hospitals."

Screening Research: Epidemiological Examples

The prevailing outlook of epidemiologists on research concerning the usefulness of screening for a cancer is most eminently exemplified by the research – both original and derivative – on screening for *breast cancer*; but most recently it has been on display from research on screening for *prostate cancer* (and lung cancer also).

In terms of this outlook, screening is generally construed as a *single test* (propos. III – 2.22), screening for breast cancer as mammography; and the test's application is taken to be a matter of a program in *community medicine* (propos. III – 2.22), in which "Persons with positive or suspicious findings must be referred to their physicians for diagnosis and necessary treatment" (ref. 1). The purpose of a program of screening for a cancer in this framework of thought is taken to be *reduction of mortality* from it (propos. III – 2.22). Thus, "The purpose of the [Malmö mammographic screening trial; ref. 2] was to assess whether [*sic*] repeated invitation to mammography reduces mortality from breast cancer." Given this type of purpose, epidemiologists view screening for a cancer as a community-level *intervention* (to reduce mortality from a cancer; propos. III – 2.22).

In respect to research on "whether" screening for a cancer, in this meaning of screening, serves this type of purpose, the core methodologic tenet of epidemiologists is this: "Owing to the potential lead time (the amount of time by which diagnosis is advanced through screening) and to length time bias associated with screening (the tendency of screening to pick up slow growing tumours) a *random-ized trial* is necessary to determine whether such a reduction [in mortality] does occur" (ref. 2; italics added).

The Abstract of the report, in 1988, on the *Malmö mammographic screening trial* (ref. 2) describes the study this way:

> *Study objective* – To determine whether mortality from breast cancer could be reduced by repeated mammographic screening.
>
> *Setting* – Screening clinic outside main hospital.
>
> *Patients* – Women aged over 45; 21088 invited for screening and 21195 in control group.
>
> *Interventions* – Women in the study group were invited to attend mammo-graphic screening at intervals of 18–24 months. Five rounds of screening were completed. . . .
> *End point* – Mortality from breast cancer.
>
> *Measurements and main results* – . . . 63 *v* 66 women died of breast cancer (no significant difference; relative risk 0.96 (95% confidence interval 0.68 to 1.35)). . . . More women in the study group died from breast cancer in the first seven years; after that the trend reversed, especially in women aged $\geq$ 55 at entry. Overall, women in the study group aged $\geq$ 55 had a 20% reduction in mortality from breast cancer (35 *v* 44; relative risk 0.79 (0.51 to 1.24)).
>
> *Other findings* – . . . Cancers classed as stages II–IV comprised 33% (190/579) of cancers in the study group and 52% (231/443) in the control group.
>
> *Conclusions* – Invitation to mammographic screening may lead to reduced mortality from breast cancer, at least in women aged 55 or over.

Some editorial notes are again in order. The study could not possibly "deter-mine" whether the screening serves its purpose; it could only constitute a test of that idea, production of evidence for inference (uncertain) about it. The participants in the study were not "patients," as they were not suffering from breast cancer. Mammography per se is not intended to change the course of a woman's health; it is a diagnostic procedure ('test'), not an "intervention." The outcome of interest ("end point" of follow-up) was death from breast cancer, not (rate of) "mortality" from it. Those statistics 0.96 and 0.79 are not "relative risks" but merely empirical counter-parts ('point estimates') of these; they are rate ratios (empirical). That in the study group there were 20% fewer deaths from breast cancer than in the control group (of women at least 55 years of age) does not mean that there was a 20% "reduction" in this the mortality (consequent to screening-provided early treatments in place of later, symptoms-associated treatments; that difference lacks statistical significance, even). That invitation to mammography "may lead" to reduced mortality from breast cancer is a possibility, not a conclusion; it presumably was recognized as a possi-bility when the study was being planned, instead of being something learned from the study. (Charmingly, perhaps to give the air of epidemiological – community-medicine – research, the "setting" for the study – for what actually was its clinical work – was a facility "outside the main hospital.")

As a matter of further need for editing, the report states that "The study was designed to document a 25% reduction in mortality in breast cancer with a power of 0.90 at the 5% level of significance." In truth, however, that designed "power" meant only that there was estimated to be a 90% probability of $P < 0.05$ if there should be a 25% reduction of risk of death from breast cancer in the screened subcohort, over the duration of follow-up, as a consequence of the 'screening' (its associated early treatments, i.e.). The 90% was not the designed probability "to document a 25% reduction" in the (average) risk. Such documentation was impossible to achieve.

It is left unspecified whether that 25% referred to a possible consequence of actual screening, or to screening with such a degree of adherence to the experimental schedule as was anticipated; but regardless, specified should have been the degree of screening that was anticipated to occur among those invited, and also to those not invited, for the screening. As it was, "The attendance rate was higher in the first round (74%) than subsequent rounds (70%) and higher among younger than older women"; and, "A random sample of 500 women in the control group showed that 25% had undergone mammography during the study period, most only once." No point is made of enforcement of the experimental assignments – even though randomization without follow-through conduces to biased results (propos. III – 4.16, 20).

Whatever may have been involved in that 'sample size determination' actually is irrelevant for the evidentiary burden of the study report (propos. III – 4.27); but the incompleteness of adherence to the screening regimen in the "study group" and the screening that took place in the "control group" meant a substantial downward bias in the measure of the mortality reduction, such as it was (cumulative mortality from breast cancer over the entire duration of the follow-up).

Like the 'sample size determination,' the reported principal result on the "reduction in mortality from breast cancer" involves *no distinction-making among different periods of time since the initiation of the screening*, as indeed continues to be customary in epidemiologists' trials – clinical (!) trials – on screening for a cancer. The report's Discussion section has, however, content at variance with this routine of acting as though the rate ratio could be presumed to be constant over time since the initiation of the screening:

> The life cycle of breast cancer is long, lasting on average about 15 years [refs.]. Accordingly, intervention at the non-invasive or early invasive stage would not influence the death rate until several years later. ... It thus is reasonable to assume that the effect of screening for breast cancer is delayed, a point that was recently considered in a review [ref.]. After a six year delay ... our study showed a 30% reduction in mortality from breast cancer; when preliminary data from 1987 are included the reduction is 42% [X% "reduction" being a misrepresentation of X% lower rate].

In 2001, thirteen years after the publication of this report, an eminent derivative study (ref. 3) addressed all reported randomized trials on screening for breast cancer, having involved a total of some half-a-million women. Only two of the trials were deemed to have been valid and, thus, contributory to the aggregate of "reliable" evidence. The Malmö study contributed the bulk of the thus-sanctioned evidence, and the review led to the claim that "there is no reliable evidence that screening for breast cancer reduces mortality" (ref. 3). The focus was, again, on cumulative incidence of death from the cancer over the entire duration of follow-up after the screening's initiation, without any regard for the duration of the screening (very short in the only other purportedly valid study) or that of the follow-up.

The epidemiological culture manifest in the 1988 report on the Malmö trial on screening for breast cancer is just as manifest in the 2009 report on the *ERSPC* (*European Randomized Study of Screening for Prostate Cancer*; ref. 4). The report's Abstract includes these elements:

> *Background* – The [ERSPC] was initiated in the early 1990s to evaluate the effect of screening with prostate-specific antigen (PSA) testing on death rates from prostate cancer.

> *Methods* – We identified 182, 000 men ... through registries in seven European countries for potential inclusion in our study. The men were randomly assigned to a group that was offered PSA screening at an average of once every 4 years or to a control group that did not receive such screening. ... The primary outcome was the rate of death from prostate cancer. ...

> *Results* – In the screening group, 82% of men accepted at least one offer of screening. During a median follow-up of 9 years, the cumulative incidence of prostate cancer was 8.2% in the screening group and 4.8% in the control group. The rate ratio for death from prostate cancer in the screening group, as compared with the control group, was 0.80 (95% confidence interval [CI], 0.65 to 0.98; adjusted $P = 0.04$). ...

> *Conclusions* – PSA-based screening reduced the rate of death from prostate cancer by 20% but was associated with a high risk of overdiagnosis.

Some editing is again called for. The "primary outcome" actually was death from prostate cancer, not the rate of this. The reported cumulative rates of incidence are not those of "prostate cancer" per se but of (the event of) diagnosis (rule-in) about prostate cancer. Scientific conclusions are not properly expressed in past tense, only experiences are; and it was mere experience that the cumulative rate of prostate-cancer diagnosis was 20% lower in the screened cohort, which is not to say that it was 20% "reduced" (by the screening-associated early treatments).

The most remarkable feature of this report is the following added passage (beyond that quoted above) in the Results section of the Abstract:

> The absolute risk difference [actually difference in mere empirical rate of death from prostate cancer over a median of nine years of follow-up] was 0.71 death

per 1000 men. This means that $[1000 / 0.71 =]$ 1410 men would need to be screened and $[(8.2\% - 4.8\%) / 0.00071 =]$ 48 additional cases of prostate cancer would need to be treated to prevent one death from prostate cancer. [That 48 is the total number, by this calculation, 47 the corresponding additional one.]

Apropos, the associated *editorial* (ref. 5) makes, uncritically, the dramatic point that "The report on the ERSPC trial appropriately [*sic*] notes that 1410 men would need to be screened and an additional 48 would need to be treated to prevent one prostate-cancer death during a 10-year [*sic*] period, assuming the point estimate is correct."

The simple truth of the matter is, however, that the screenings and treatments cannot realistically be expected to fully – or even mainly – deliver their life-saving benefit within that typical period of follow-up from the screening's inception, nor does the mortality reduction begin as of the inception of the screening. The ERSPC report itself makes the observation that "The rates of death in the two study groups began to diverge after 7 to 8 years and continued to diverge further over time (Fig. 2)." Yet in total disregard of this evidence, even, the first point in the Discussion is the putative need for 1410 screenings and 48 additional treatments "to prevent one prostate-cancer death." The Discussion shows no awareness of the delay in the manifestation of the mortality reduction that may result from the introduction of screening (in a cohort-type population).

The "conclusion" that the screening "was associated with a high risk of over-diagnosis," it seems, was based on the completely spurious idea (cf. above) that only one out of 49 early treatments provided for by screening actually cures an otherwise fatal case of prostate cancer – so that 48 of 49 cases represent overdiagnosis and its consequent overtreatment.

As should be apparent, epidemiologists' current ideas about research on screening for a cancer are empiricist to the point of conspicuous absence of the "mental legislation" that should be "founded on the nature of reason and the objects of its exercise" (propos. I – 2.9).

Each of the two epidemiological studies on screening for a cancer that are addressed above (refs. 2, 4) has been revisited from the outside in an attempt to introduce some rationality into the way evidence from such studies is used to quantify the mortality implications of the screening (refs. 6, 7).

These outside commentaries have sprung from the fundamental recognition that *nothing meaningful is being quantified* by the ratio of mortality from the cancer – contrasting a screened cohort with an unscreened one – over the entire, *arbitrary* duration of follow-up as of entry into the trial, this in the context of an *arbitrary* – generally quite short – duration of the screening in the trial. In particular, that ratio does not serve as a measure of how much mortality from the cancer can be expected to be reduced if a given pattern of the screening – pursuit of the cancer's early rule-in diagnosis (propos. II – 1.24, III – 2.20–22) – has been prevailing in a particular population for a duration long enough for the reduction to have fully materialized; nor does it serve to quantify the reduction in the cancer's case-fatality rate resulting from its early treatments.

The commentary (ref. 6) on the Malmö trial (ref. 2) takes as its point of departure the Malmö investigators' point that "It is thus reasonable to assume that the effect of screening for breast cancer is delayed. ... After a six year delay ... our study showed a 30% [lower] in mortality from breast cancer; when preliminary data from [one more year of study] are included the [corresponding result] is 40%." And the commentary goes on to show that, with focus on time at least eight years subsequent to the screening's initiation the mortality-rate ratio was as low as 0.45 with a 95% (im)precision interval from 0.24 to 0.84 – this in the context of only about 70% adherence to the experimental screening, scheduled to be repeated at intervals of 18–24 months, and with a proportion as high as 25% of the control cohort also screened (at least once). (Without this focus, as noted above, the Malmö investigators reported a rate ratio of 0.79 with a 95% interval from 0.51 to 1.24.)

The commentary presents the rationale for the clinical interpretation of this substitute result – of the complement of this suitably time-specific rate ratio in the context of suitably long-term screening – as the empirical value for the *proportional reduction in the case-fatality/incurability rate* of breast cancer resulting from the screening, from screening-associated early treatment in place of what otherwise would be symptoms-associated late treatment. In the Malmö study, the screening had continued long enough for this reduction to be manifest to an appreciable extent, though by no means fully. For, that commentary makes the point that a randomized trial on screening for a cancer provides for manifestation of the proportional reduction in the cancer's incurability rate – in a particular segment of the study cohort's follow-up time – if and only if the duration of the screening exceeds the difference between the maximum and minimum of the times from screening-provided cures of the cancer to the deaths that are thereby averted (in otherwise fatal cases of the cancer). If that maximum and minimum are T_2 and T_1, respectively, and S is the duration of screening, the quantitatively relevant segment of follow-up time is from T_2 to $S+T_1$ (focus on which requires that $S+T_1 > T_2$; i.e., that $S > T_2-T_1$). With $T_1 = 7$ yr. and $T_2 = 20$ yr., the full reduction in the incidence density of death from breast cancer in that type of trial, if valid, would begin as of 20 yrs. of follow-up if the screening continued for at least 13 yrs.

That substitute result from a sufficiently long-term valid trial of this type has a much more subtle and vague interpretation for the mortality that is the concern in epidemiology (in community medicine, i.e.). It serves as an estimate of the degree to which mortality from the cancer would be reduced (on account of screening) if everyone who in the absence of any screening would be dying from the cancer had, at that time, a history of having been screened (according to the study protocol) throughout the period from T_2 ago to T_1 ago. This constitutes a basis for surmising – with considerable uncertainty – how much the mortality would be reduced – in the fullness of time – if that condition for the maximal possible reduction would be satisfied to a given incomplete extent. On the other hand, the kind of mortality ratio that is epidemiologists' current concern (with arbitrary features in critical respects; cf. above) is no basis for rational prognosis in community medicine (any more than in clinical medicine).

A similar re-examination of data from the European trial on PSA screening for prostate cancer (refs. 4, 7) produced a statistically highly significant rate ratio well below 0.50. (Reported by the investigators of that study was, as noted above, rate ratio 0.80 with 95% interval from 0.65 to 0.98.)

References

1. Porta M (Editor), Greenland S, Last JM (Associate Editors). A Dictionary of Epidemiology. A Handbook Sponsored by the I. E. A. Fifth edition. Oxford: Oxford University Press, 2008.

2. Andersson I, Aspergren K, Janzon L, et alii. Mammographic screening and mortality from breast cancer: the Malmö mammographic screening trial. BMJ 1988; 297: 943–8.

3. Olsen O, Gøtzsche PC. Cochrane review of screening for breast cancer with mammography. Lancet 2001; 358: 1340–2.

4. Schröder FH, Hugosson J, Roobol MJ, et alii. Screening and prostate-cancer mortality in a randomized European study. NEJM 2009; 360: 1320–8.

5. Barry MJ. Screening for prostate cancer – the controversy that refuses to go away. Editorial. NEJM 2009; 360: 1351–4.

6. Miettinen OS, Henschke CI, Pasmantier MW, et alii. Mammographic screening: no reliable supporting evidence? Lancet 2002; 359: 404–6; image.thelancet.com/extras/1093web.pdf.

7. Hanley JA. Mortality reductions produced by sustained prostate cancer screening have been underestimated. J Med Screen 2010; 00: 1–5.

Screening Research: A Clinical Program

In 1992, D. B. Skinner as the then head of thoracic surgery at the New York Presbyterian Hospital of the Cornell University Medical Center (in New York City) set out to identify and put in motion a line of research that would engage the various disciplines concerned with thoracic medicine there – the pulmonologists, the chest radiologists, the pathologists, the oncologists, and, of course, the thoracic surgeons. So, a 'retreat' was held (in a ski resort in Utah). In it, colleagues from each of those disciplines outlined their research concerns, and the teacher of this course was an invited attendant with a view to making a recommendation on what the common line of research might be.

The recommendation was *CT screening for lung cancer*, and this recommendation was supplemented with a broad outline of the nature – *clinical, multidisciplinary* (propos. II – 1.24) – of the possible program of research. The recommendation was readily and unanimously accepted, and this led to a more detailed plan for the research and then to the initiation of the *Early Lung Cancer Action Program, ELCAP*. This program has grown into a still on-going major international collaboration, the International ELCAP, I-ELCAP (ref. 1). In 2010 it held the 23rd semiannual international conferences on CT screening for lung cancer with these conferences scheduled to continue in 2011 and beyond.

From the clinical vantage of ELCAP, screening for a cancer is *not* regarded as community-setting application of a test (or more than just one test) to asymptomatic people with the idea of referring those with a positive result of this testing for (possible rule-in) diagnosis and its consequent early treatment in a clinical setting; nor is it viewed as a community-level – or clinical – intervention with the purpose of reducing (the rate of) mortality from the cancer in a community.

From the clinical vantage of ELCAP, screening for a cancer is the pursuit of the cancer's early, latent-stage detection (rule-in diagnosis) with a view to thereby being able to take advantage of the greater effectiveness of – more common curability by – early treatment in comparison with treatment when the cancer already is clinically manifest (propos. III – 2.20). It is viewed as application of a particular *diagnostic regimen* (propos. III – 2.21), an algorithm that only begins with a single test. If this test's result is negative (as defined for the purpose), the pursuit stops; but if it is positive, the pursuit continues – possibly all the way to satisfying the final, pathological criterion for rule-in diagnosis about the cancer.

From this clinical conception of the essence of screening for a cancer, adopted as a novel one in ELCAP, has flown the understanding that a necessary prerequisite for any actual research on screening for a cancer is development of the regimen of screening – rather aprioristic development of the first version of this; and then, as experience with this version accrues, updating of this regimen; etc. For example, as for the initial CT test in baseline screening, ELCAP experience allowed restriction of its positive result regarding solid nodules (non-calcified) to ones at least 5 mm in diameter, thereby substantially reducing the frequency of the result's consequent diagnostic work-up without any apparent reduction in the frequency of cancer diagnosis before the next repeat screening (12 months later).

Given what, at any given time, has been taken to be the screening regimen of choice, the first-order clinical-research concern about it in the ELCAP has been (and still is) its diagnostic performance properties in a given round of the screening, distinguishing between baseline and repeat screening. The objects of study in this context have been two: the distribution of the diagnosed cases (screen- and interim-diagnosed cases combined) according to prognostic indicators (most notably stage and stage-conditional size); and the probability that a round of the screening will lead to diagnosis (rule-in) about the cancer. The former has been presumed to be essentially independent of indicators of risk for the cancer, while the latter has been studied as a function of these indicators (incl. time since the most recent screening/imaging, the result of its first test having been negative).

Even though screening for a cancer is viewed as a diagnostic pursuit from the clinical vantage of the ELCAP, calling for *diagnostic research*, this has not been the entire concern in the ELCAP, as the screening conduces to novel prognostic challenges. An eminent example of this has been the high proportion, among the cases diagnosed at baseline, of cancers that in the imaging present as 'nodules' that evidently are not solid, the appearance being that of 'ground glass opacity' instead. There are good reasons to believe that these 'subsolid' ('part solid' and 'non-solid') cancers are distinctly more slowly growing than the solid ones, and this raises the possibility of *overdiagnosis* – and in any case *overtreatment* – in these cases. The concept of overdiagnosis in ELCAP is the natural one: diagnosing as malignant – as cancer – a lesion that actually is benign (i.e., does not represent uncontrolled neoplasia). Treatment of an overdiagnosed case of cancer obviously represents overtreatment; but so also does treatment of a particularly slowly grow-ing genuine cancer, especially in the context of an otherwise relatively short life expectancy.

This has, in the ELCAP, meant that the purely diagnostic research on any given regimen of CT screening for lung cancer needs to be supplemented by *prognos-tic research* into the prospective course especially of these novel types of the cancer diagnosed – possibly overdiagnosed – by the screening, including course when treat-ment is delayed under 'watchful waiting' (if not completely withheld). Among the objects of the prognostic research are the proportion of overdiagnosed cases among cases screen-diagnosed as Stage I cancers, the proportion of curable cases among the diagnosed ones, by stage, and the distribution of the time lag from early treatment of genuine cases and the death that thereby is prevented.

The methodology of the diagnostic research on a regimen of screening for a cancer – the screening being a consideration in the context of 'the worried well' concerned about a particular cancer – has in the ELCAP been understood to be very different from the methodology of the diagnostic research that advances the scientific knowledge-base of pursuing diagnosis in the context of patients presenting with a particular complaint. The methodologic difference flows from the difference in the form of the respective objects of study.

In ELCAP, asymptomatic persons at relatively high (near-term) risk for (an overt case of) lung cancer – from the domain of potential screening in practice – are recruited for experimental screening (following informed consent), baseline screening and (thus far mainly) annual repeat screenings. Upon documentation of recent absence chest imaging, lack of symptoms of lung cancer and status in respect to risk indicators for lung cancer (as well as fitness to undergo thoraco-tomy), baseline screening is carried out. Of the accrued series of these baseline screenings, the subseries that has led to diagnosis (rule-in) about lung cancer (fol-lowing the program's diagnostic protocol) is the basis of the result in respect to the diagnostic distribution at baseline by prognostic indicators (most notably stage and stage-conditional size); and the series of baseline screenings as a whole provides for producing a function (logistic) expressing the baseline prevalence of (lesions diagnosed as) lung cancer as a joint function of the indicators of (near-term) risk for (an overt case of) lung cancer (notably age and history of smoking). The diag-nostic research pertaining to the repeat screenings is, naturally, wholly analogous to this research on the regimen's baseline variant (though with time since the previ-ous screening – with its result accounted for – now an important determinant in the prevalence function).

Regarding the prognostic implications of Stage I screen-diagnosed cases, I-ELCAP has, for one, addressed their rate of *curability* on the premise – question-able – that no overdiagnosis has been occurring. In this, the curability rate has been assessed as the asymptotic value of the 'cause-specific' survival rate (in which lung cancer is the only cause of death; refs. 2–3). For another, I-ELCAP has spawned a randomized trial on the treatment of these cases, which provides for addressing the frequency of overdiagnosis as well as curability of genuine cases of the cancer (ref. 4).

References
1. www.IELCAP.org
2. The International Early Lung Cancer Action Program Investigators. Survival of patients with Stage I lung cancer detected on CT screening. NEJM 2006; 355: 1763–71.
3. Miettinen OS. Survival analysis: up from Kaplan-Meier-Greenwood. Eur J Epidemiol 2008; 23: 585–91.
4. Phase III randomized study of lobectomy versus sublobar resection in patients with small peripheral stage I non-small cell lung cancer. National Cancer Institute. Clinical trials (PDQ). www.cancer.gov.

PART V
EPILOGUE ON MAJOR IMPROVEMENTS IN CLINICAL MEDICINE

V – 1. THE PREDICATES OF MAJOR IMPROVEMENTS
V – 2. DEVELOPMENT OF THE MAJOR IMPROVEMENTS
V – 3. RESEARCH FOR FURTHER IMPROVEMENTS
V – 4. 'CLINICAL EPIDEMIOLOGY' & EBM AS SET-BACKS

V – 1. THE PREDICATES OF MAJOR IMPROVEMENTS

This course was fundamentally predicated on medicine in the West now being a vast and highly technological industry imbedded in a 'culture of improvement,' with recent dreams of a central role for information technology: that the knowledge-base of clinical medicine would get to be codified in, and available for ad-hoc retrievals from, practice-guiding *expert systems* (Preface).

To this fundamental predicate was joined a supplementary one: that, just recently, "theoretical progress has produced understanding of the *forms* in which the knowledge-base of clinical medicine should be codified, and of the way its *content*, in terms of those forms, can and must be garnered from clinical experts' tacit knowledge"; in other words, that "the requirements now are in place for the development of practice-guiding expert systems . . . for truly Information-Age practice of clinical medicine" (Preface).

To this latter predicate was attached the remark that, "By the same token, it now is clear what types of knowledge is to be pursued in clinical research to make the systems ever more scientific in their content." (Preface).

The fundamental predicate (above) was given an alternative expression in proposition II – 3.2: "A modification of Cochrane's premise [that knowledge about effectiveness, from clinical trials, serves to enhance the efficiency – and hence cost-effectiveness – of healthcare] deserves consideration: If doctors were able to know, right in the course of their practices, in respect to the type of situation that confronts them at a given moment, what their most illustrious colleagues in the same situation typically do (as a matter of fact-finding) and think (as a matter of translating the available facts into the corresponding gnosis [diagnosis, say]), they would tend to do and think likewise. Thus, if it be possible for doctors to know this, a consequence would be an increase in the most productive – cost-effective – testings and interventions and a corresponding reduction in relatively wasteful ones. In this Information Age the implication is that the availability of user-friendly gnostic *expert systems* would enhance the efficiency of healthcare by inherently contributing to both quality assurance and cost containment in it." This fundamental premise, it should be noted, has its meaning in a 'culture of improvement' rather irrespective of the extent to which experts' tacit knowledge has been advanced by clinical research. It is a matter of *development* without any inherent connection to research.

O. S. Miettinen, *Up from CLINICAL EPIDEMIOLOGY & EBM*,
DOI 10.1007/978-90-481-9501-5_15, © Springer Science+Business Media B.V. 2011

Regarding the forms in which the knowledge-base of clinical medicine should be codified, this course presented as an early exemplar a study by Pozen et alii for diagnosis about myocardial ischemia, a quarter-century ago. The didactic point about this study principally was that it addressed, for myocardial ischemia in a suitably defined domain of patient presentation, the diagnostic probability as a joint function of the set of diagnostic indicators relevant and available in the context of the decision at issue (about admission into CCU); and an added point was that this function was made user-friendly by means of appropriate technology (programming into a hand-held calculator).

In line with this paradigm, a central point of this course (under Theory of Clinical Medicine) was proposition II – 2.12: "For the knowledge-base of clinical gnosis [dia-, etio-, prognosis], the necessary (only practical) form – so long as the relevant distinctions (propos. II – 2.1–3) are being made – is that of *occurrence relations* (ref.), formulated as empirical models for the probabilities. For, focus on these *gnostic probability functions*, GPFs, commonly reduces the need to know, separately, about an enormous multiplicity (thousands) of probabilities for a given object of gnosis in a given domain, to the need to know about the magnitudes of the very much smaller number (at most dozens) of parameters involved in a reasonable model that addresses all of those probabilities."

Concerning the predicate that there recently has been the requisite progress to understand how experts' tacit, gnosis-relevant knowledge can and should be garnered in terms of GPFs, among the core points of the course was proposition II – 3.15: "Expert clinicians' gnosis-relevant general knowledge is not something they could make explicit in the form of GPFs or in some other general terms. Their knowledge is *tacit* in nature. They know about gnostic probabilities only ad hoc, in practice when gnostic challenges present themselves in their clinical encounters with clients; and in these instances, even, in terms that are inconsistent across individual experts. Thus *the challenge* is to garner experts' tacit knowledge in the form of their typical ad-hoc beliefs about probabilities (propos. II – 2.11) and to give the pattern of these the form of GPFs – this on the premise that expertise on the topic actually exists."

This was supplemented by proposition II – 3.16: "Given that expert clinicians know about gnostic probabilities in instances of gnostic challenge that actually occur in their practices, it follows that they equally know about them in hypothetical instances. From this it follows that insofar as experts' tacit knowledge about gnostic probabilities – for a particular object in a particular domain – exists, garnering it is most efficiently done on the basis of *hypothetical* instances presented to them; and the developmental challenge thus reduces to giving the thus-garnered tacit knowledge the form of a GPF addressing experts' typical beliefs."

V – 2. DEVELOPMENT OF
THE MAJOR IMPROVEMENTS

Those predicates inspired proposition II – 3.7: "The dream of *universal excellence* in clinical medicine can be expressed this way: When a person consults a doctor (in the relevant discipline of clinical medicine), it does not matter who the doctor is: rather than a creative thinker subject to 'cognitive errors' (ref. 1), the doctor inherently represents to the client access to – interface with – the knowledge that characterizes the top experts in the discipline … and gives, in his/her teaching … the client the full benefit of this expertise." Importantly, the predicates (above) imply that this is not just an idle, utopistic dream; to repeat: "the requirements now are in place for the development of practice-guiding expert systems" as the basis for what truly is Information-Age type of practice of clinical medicine.

Proposition II – 3.8, apropos, quotes Goethe as having deduced from cognizance of "what we are capable of" a point that can be regarded as the psychological basis of the operational upshot of our 'culture of improvement': "Passionate anticipation thus changes that which is materially possible into dreamed reality." The question thus is about the extent to which there are, or will be, *leaders* of the various disciplines of clinical medicine who are gripped by this "passionate anticipation" of universal excellence in (the practice of) clinical medicine.

This course was intended to enhance the prospects in this regard: "The overarching aim of this course was to sow seeds of *major improvements* in clinical medicine in this Information Age." And so, "One specific aim of this course thus was to orient some of the students … to the path through which they would become maximally *productive leaders*, and thereby agents of major improvements, in their respective disciplines ('specialties') of clinical medicine, now that the era of genuinely scientific medicine – its theoretical framework rational and its knowledge-base from science (ref. 1) – is dawning (ref. 2)." (Cf. Aims of the Course.) The major improvements whose agents the leaders were seen to become were matters of the advent of universal expertise in the various disciplines of clinical medicine, resulting from the availability and common deployment of gnostic expert systems (cf. above).

Pioneering work towards this dreamt-of reality – methodologic development and demonstration projects – is already underway in the Horten Center for Patient-Oriented Research and Knowledge Transfer of the University of Zurich, under the direction of J. Steurer (here the author of the Foreword). That work should turn out to have been the precursor of a pan-European network of discipline-specific programs of the production of GPFs and expert systems based on these; and this, in turn, should inspire the formation of its North American counterpart, among others – including ones specific for various developing-country settings. But if the Western 'culture of improvement' (Preface) turns out to be slow to take up the mission, perhaps China will be more progressive – thus again reminding us in the West of the Roman adage, *Ex Oriente lux*.

This vision of the Western 'culture of improvement' in action in Information-Age medicine the course presented under the broad rubric of Theory of Clinical Medicine, and not under Theory of Clinical Research. For it represents a vision of what may be only *quasi-scientific medicine*, as envisioned is medicine which, while having a rational theoretical framework (akin to truly scientific medicine), can have a knowledge-base that merely resembles scientific knowledge (about GPFs), without the knowledge actually being a product of science. The vision is one of medicine at large being rational – including knowledge-based – in this sense.

As noted above, this vision's translation into reality is a matter of mere *development*, rather than of research followed by development. But it is a matter of development bringing about very *major improvements*, without correspondingly major investment.

V – 3. RESEARCH FOR FURTHER IMPROVEMENTS

Given what clinical medicine is – the aggregate of the arts/disciplines of (the practice of) clinical healthcare and not a (set of) science(s) (propos. II – 1.7, 8) – there is no science *in*, or *of*, clinical medicine; but there is, of course, science *for* clinical medicine, ultimately quintessentially 'applied' clinical science for the advancement of the knowledge-base of clinical medicine (propos. I – 2.5) – providing for clinical medicine that is not only *rational* in its theoretical framework (addressed above) but increasingly *scientific* as to the genesis of its knowledge-base.

A major theme in this course – in its Theory of Clinical Medicine part, preparatory to the Theory of Clinical Research part – was the *necessary form of the knowledge-base* of clinical medicine, namely that of GPFs (cf. above), along with the point that the use of GPFs can be made practical by means of their incorporation into gnostic expert systems. This has a critically important bearing on *objects design* in quintessentially 'applied' clinical research and, thereby, on the *form of the results* – of the numerical evidence – that such research is to produce.

A related major theme was the necessary movement from the evidence produced by research with appropriate objects design to *knowledge* of the form of the objects of research – generally evidence from derivative rather than original research. While the making of these transitions was presented as being, in principle, a function of the relevant, topic-specific scientific communities, the operational proposition was that the relevant evidence be made – suitably – to enhance the tacit knowledge of the members of the various *expert panels* that are the source of the GPFs for the expert systems (propos. III – 1.15–17).

The students in this course – residents and fellows in the McGill University Health Centre – reviewed nine example studies of their own, collective choosing, presenting them in class during the last (fourth) week of the full-time course. They judged none of them to provide suitable guidance for their practices, even if the results be taken to represent actual knowledge about the respective objects of study. Throughout, the problem in this was the *form* of the result, that it was too simplistic, not providing for the necessary distinctions that are to be made. They weren't suitably-designed GPFs in form.

O. S. Miettinen, *Up from CLINICAL EPIDEMIOLOGY & EBM*,
DOI 10.1007/978-90-481-9501-5_17, © Springer Science+Business Media B.V. 2011

V.2 RESEARCH FOR
FURTHER IMPROVEMENTS

V – 4. 'CLINICAL EPIDEMIOLOGY' & EBM AS SET-BACKS

This course was a regularly scheduled one on "clinical epidemiology" – this time one with a 'guest' teacher. As 'clinical epidemiology' purportedly is "a basic science for clinical medicine" (ref. 1) and was spearheaded by clinicians who on this basis became leaders of the EBM (Evidence-Based Medicine) movement (ref. 2), the course naturally also was about EBM. For it is not reasonable to teach the means without regard for the end. And very importantly, as for both the end and the means, it is not reasonable to teach them as though the contents put forward by those leaders were canonized truths just because their following in medical academia has so swiftly grown so large (propos. I – 5.15): the responsibility of a tenured professor in particular is to teach the *truth*, as best (s)he sees it, for the students to "weigh and consider" (propos. I – 1.1) – heeding "the praiseworthy saying of Socrates: 'But no man must be honored before the truth' " (ref. 3).

References:
1. Sackett DL, Haynes RB, Gyatt GH, Tugwell P. Clinical Epidemiology. A Basic Science for Clinical Medicine. Second edition. Boston: Little, Brown and Company, 1991.
2. Sackett DL, Straus SE, Richardson WS, et alii. Evidence-Based Medicine. How to Practice and Teach EBM. Second edition. Edinburgh: Churchill Livingstone, 2000.
3. O'Malley JW. Four Cultures of the West. Cambridge (MA): The Belknap Press of Harvard University Press, 2004; p. 79.

It is instructive, for orientation here, to take note of the *seminal event* in the genesis of 'clinical epidemiology' and EBM: "it dawned on [D. L. Sackett] that epidemiology and biostatistics could be made as relevant to clinical medicine as his research into the tubular transport of amino acids" (props. I – 5.3, IV – 1.2). Meant by this envisioned equivalence presumably was that those two disciplines could be made very (*sic*) relevant to clinical medicine, while in their then forms they weren't very relevant (different from the nephrophysiological research DLS was conducting); and presumably meant also was that Sackett and his clinician colleagues *themselves* could effect this envisioned development – despite their having no record at all in the advancement of the theory of epidemiology, or biostatistics, to the then states of these.

In the end, as it turns out, the vision was even more ambitious: once Sackett and his colleagues will have made epidemiology and biostatistics very relevant to clinical medicine – by the creation of 'clinical epidemiology' – doctors at large should learn this new "basic science of clinical medicine" (cf. ref. 1 above) and in their practices judge research evidence accordingly, in disregard of what experts may be saying (propos. I – 5.5–6). But it actually was a vision that amounted to replacing clinical professionalism by conceited dilettantism in the practice of pseudo-scientific medicine (propos. I – 2.8–9, I – 5.8, 14).

In a way that vision has materialized, however. Despite the obvious impracticality of the founding doctrine of the EMB movement (propos. I – 5.5–6), the generally false teachings of the movement's leaders (section IV – 1), and the quite wanting understanding of clinical research by text-book authors (sect. I – 3), by editors of medical journals (propos. III – 4.26–27), and by eminent researchers themselves (sect. IV – 2), 'all clinical epidemiologists' and EBM teachings are well-established not only in Canada – their country of origin – but also in other English-speaking countries, in the main. They have come to be accepted as normal features of medical academia (propos. I – 2.8), even if rarely heeded in the practice of clinical medicine.

But: "When people accept futility and the absurd as normal, the culture is decadent" (propos. I – 2.4); and the epidemic of this new form of cultural decadence in medical academia, following the endemic that resulted from the Flexner report (propos. I – 2.8), obviously constitutes another major set-back in the pursuit of the realization of the dream of reason (propos. II – 3.7) and, thus, of major improvements in clinical medicine.

This course obviously represented rebellion against the common teachings about 'clinical epidemiology' and EBM. But "there is no contradiction between a rebellious spirit and an uncompromising pursuit of excellence in a rigorous intellectual discipline. In the history of science, it has often happened that rebellion and professional competence went hand in hand" (ref.). And should the teachings in this course, upon their necessary weighing and considering (propos. I – 1.1), be deemed not to have gone hand in hand with the requisite degree of professional competence, those of greater competence should be preparing to rebel, again in a way that is "not impulsive but carefully thought out over many years" (ref.). For, 'clinical epidemiology' and EBM, like Flexner's precursor to these, definitely represent serious misunderstanding of the essence of scientific medicine (propos. I – 2.9).

"The common element in [scientific visions] is rebellion against the restrictions implied by the locally prevailing culture" (ref.), and the common culture of today's medical academia (sect. I – 2) really should provoke rebellion by all genuine visionaries able to see how this culture restricts medical science in its noble mission to advance the arts of medicine – its clinical arts in particular.

Reference: Dyson F. The Scientist as a Rebel. New York: New York Review of Books, 2008; pp. ix, xv.

PART VI
APPENDICES

APPENDIX – 1. SOME ELEMENTARY CONCEPTS OF MEDICINE,
ACCORDING TO THE STUDENTS
APPENDIX – 2. ON THE STUDENTS' CONCEPTS, THE TEACHER'S
COMMENTS
APPENDIX – 3. ASSIGNMENTS TO THE STUDENTS
APPENDIX – 4. TO THE ASSIGNMENTS, THE TEACHER'S RESPONSES
APPENDIX – 5. MORE ON GARNERING EXPERTS' TACIT KNOWLEDGE
APPENDIX – 6. AN INDUSTRIAL PERSPECTIVE

APPENDIX – 1. SOME ELEMENTARY CONCEPTS OF MEDICINE, ACCORDING TO THE STUDENTS

In this course, scheduled to address "clinical epidemiology," the students – all of them clinical residents or fellows in McGill University Health Centre – upon having been introduced to the concept of concept and to the definition of a concept (propos. II – 1.1), were presented with a sheet of paper with a quote on the top:

> Where the concepts are firm, clear and generally accepted, and the methods of reasoning are agreed between men ... , there and only there is it possible to construct a science ...
>
> *Reference*: Berlin I. The Proper Study of Mankind. An Anthology of Essays (Hardy H, Hausheer R, Editors). London: Chatto and Windus, 1997; p. 61.

Below this quote (and reference) was this request: "*Define* (succinctly) each of the following: medicine and clinical medicine; sickness, symptom, and syndrome; illness and disease; diagnosis and prognosis; good diagnosis and good prognosis; and accuracy of a diagnostic test."

In a random sample of 10 from among the returns, the definitions were these:

Medicine

1. Applying knowledge to diagnose and treat diseases to the best care possible.
2. General term applying to the study, treatment, and care of humans with disease.
3. It is a science [of how to treat patients].
4. The act of healing.
5. Study of health and diseases.
6. The study of the human body, its variations & its illnesses & its treatments.
7. The study of illness in the human body.
8. Study of health and disease.
9. Study of diseases and their treatment/evolution.
10. The practice of identifying and treating diseases of the human condition.

Clinical medicine

1. Medicine that is applied when directly involving a patient.
2. Applies specifically to the use of medical concepts and principles by the physician while interacting with the patient.

3. The application of medicine in clinical practecles [*sic*].
4. The science of applying knowledge to treat.
5. Application of medical science to patients.
6. Application of knowledge of the human body & its conditions.
7. The use of human senses to study illness in the human body.
8. Direct interaction with patients.
9. Art of treating, preventing or rehabilitating illnesses in patient[s].
10. The practice of identifying and treating disease in the context of the patient-doctor relationship.

Sickness

1. The absence of health in one or more systems of the body.
2. A state of unwellness experienced by the patient.
3. Unwell, or disturbe [*sic*] the N physiology.
4. To be ill.
5. The subjective feeling of being sick, i.e., not in the usual state of health.
6. Occurrence when the human body does not function as it should or functions as it should not.
7. A state of being that is universally accepted as being unhealthy.
8. -------
9. Change from usual state of health.
10. A disruption of the normal homeostasis mechanisms of the human body.

Symptom

1. Appearance or presentation of a condition that is not healthy / in accordance with a normal functioning body.
2. A bodily manifestation of disease experienced by the patient.
3. Manifestation of the disease (presentation of the disease).
4. The appearance of changes due to a disease.
5. Clinical manifestation of an [*sic*] health problem that is experienced by a patient.
6. Manifestation of a sickness.
7. A subjective interpretation one has for their illness.
8. Patient-reported feeling or condition.
9. An [*sic*] subjective feeling by a person that is attributed to an underlying sickness.'
10. Element that is qualitative and brought about by patient.

Syndrome

1. Set of symptoms that are specific for one disease.
2. Collection of signs and symptoms seen with a particular disease.

3. Multi-system disease (involvement of more than 2 disease [*sic*] with specific characteristics).
4. Collection of symptoms.
5. Cluster of symptoms or signs or malformations that can happen together in different individuals.
6. Collection of symptoms.
7. A collection of symptoms comprising an overall state of health.
8. A collection of symptoms.
9. Group of symptoms that together are a clinical entity.
10. A cluster of symptoms and/or signs that are related together and are linked to a specific condition/disease.

Illness

1. A subjective perception by a patient of an objectively defined disease (Wiki).
2. A state of unwellness experienced by the patient.
3. It could be organic or non-organic distubtion [*sic*] or defection [*sic*] of the normal physiology.
4. To be abnormal.
5. Condition that is happening in a patient but for which we don't have an explanation.
6. A state of being outside one's usual health.
7. An individual's perception of a disease.
8. Illness encompasses the disease within an individual & how he/she is affected by it.
9. The consequences of a disease on the body.
10. A feeling of unwellness that persists over time.

Disease

1. -------
2. Lacking wellness or disordered functions of physiological systems.
3. It is pathological description of abnormal physiology which end [*sic*] to abnormal manifestations (symptoms).
4. Change or deviation from normal.
5. Condition for which we have an understanding of the causes and processes that lead to the clinical manifestation.
6. -------
7. -------
8. Pathologic/pathophysiologic description of a sickness.
9. Diagnostic entity.
10. A condition that leads to disruption of a person on biological/psychological/social levels.

Diagnosis

1. To "look through" the condition = naming it.
2. Identification of a particular set of signs & symptoms.
3. Identification of disease manifestations.
4. Finding the reason of illness.
5. Determining the condition that is responsible for the patient's symptoms.
6. Defining an [sic] sickness based on a constellation of symptoms +/– laboratory tests +/– physical exam.
7. A proposed disease of a patient.
8. Identification of a disease.
9. Meeting diagnostic criteria of a disease.
10. An assessment of a patient's subjective presentation and objective analysis that leads to identification of the underlying pathology.

Prognosis

1. To look ahead of the condition's evolvement.
2. The predicted outcome of a disease.
3. Outcome or result of the disease itself or of the treatment.
4. Outcome of the disease.
5. Prediction on the course of the condition affecting a patient.
6. Predicting the course of a sickness.
7. A predicted outcome of a patient's diagnosis.
8. Expected outcome of a disease.
9. Usual course of the illness.
10. An objective assessment of the natural progression of a disease with or without a treatment.

Good diagnosis

1. To rule out other specific diseases as best as possible.
2. A disease with benign consequences.
3. It is how can [sic] a disease can be identified by investigations a [sic] manifestations.
4. Precise and correct reason of illness.
5. ?
6. Appropriate diagnosis for a given set of findings (symptoms, tests, physical exam).
7. A [sic] accurate.
8. "Good" is a value judgment dependent on who is making it and what the circumstances are.
9. Diagnosis that meets all the criteria of a disease.
10. A diagnosis that suitably explains a patient's signs and symptoms and objective assessment.

Good prognosis

1. To have a favorable outcome.
2. Recovery to a state of wellness is very likely.
3. Good outcome or result of the treatment of disease or nature of the disease itself.
4. Favorable outcome.
5. Expression generally used when a doctor judge [*sic*] that the disease course or outcome will be relatively positive.
6. Non life threatening + non debilitating.
7. A correctly predicted outcome that has left no negative consequences on a patient's current and future status.
8. "Good" is a value judgment dependent on who is making it and what the circumstances are.
9. Illness w/ a prognosis that allows return to baseline health state.
10. A prognosis that carries with it a good response to treatment or a benign progression of the disease.

Accuracy of a diagnostic test

1. Has a high sensitivity and specificity.
2. The specificity of a test to identify type of disease.
3. How the test [is?] helping to diagnosis [*sic*] the disease.
4. The probability that a test will pick up a disease.
5. Predictivity toward a particular condition.
6. How sensitive and specific a test is for a given diagnosis.
7. -------
8. How often a test is correct. Expressed in terms of sensitivity and specificity.
9. High sensitivity and specificity for the test, with very few false negative and false positive.

APPENDIX – 2. ON THE STUDENTS' CONCEPTS, THE TEACHER'S COMMENTS

The Big Picture

Apropos of the quote in the beginning of Appendix 1, Isaiah Berlin – that giant among the humanist intellectuals of the 20th century – presumably would have taken the students' definitions of the select elementary concepts of medicine, a sample of which is given in that Appendix, to imply that it is not yet possible to construct clinical science (within the disciplines of clinical medicine). But there is hope: the students were quite ready to weigh and consider the instructor's definitions of those concepts.

Medicine, Clinical Medicine

Most of the 10 definitions of medicine specify treatment as being an element in the essence of medicine, even though the students were taught that the definitional essence of a thing is something which is true of each instance of the thing (and unique to it; cf. propos. II – 1.1), and even though treatment is very exceptional, rather than routine, in the practice of medicine (as they were going to be taught; propos. II – 1.7).

Most of the 10 definitions specify "study" as being in the essence of medicine, possibly meaning – correctly – fact-finding toward gnosis about the 'health' of the client (propos. II – 1.7); but there is lack of understanding as to what the gnosis generally is about: it is said to be about disease in six of the 10, illness in only two (cf. propos. II – 1.6).

None of the 10 definitions of *clinical* medicine expressly specifies it – within the proximate genus of medicine – as being concerned with individual clients, one at a time, thus distinguishing it from community medicine – epidemiology, that is (with a population as the client).

Sickness, Symptom, Syndrome

Several of the 10 definitions specify sickness as a state of ill-health, failing to appreciate that motion sickness, morning sickness, etc., are phenomena of unwellness in a state of perfect health (but excessive circumstantial stress; cf. propos. II – 1.6), that sickness is not a manifestation (overt) of illness alone.

Symptom – an entirely subjective manifestation of an illness – is, remarkably, misrepresented by almost all of the 10 definitions.

Syndrome – a particular cluster of symptoms and/or clinical signs that is definitional to an illness (perhaps because the somatic anomaly remains unknown) – is misrepresented by most of the 10 definitions.

Illness, Disease

'Illness' should be understood to refer to *any* ill-health, as does *maladie* in French and *Krankheit* in German; to a somatic anomaly either manifest in sickness or with the potential to so manifest. (Ref. with propos. II – 1.6.) None of the 10 definitions – highly varied – reflects this understanding.

Disease – so eminent in the 10 definitions of medicine – none of the 10 definitions identifies a process-type illness (L. *morbus*), as distinct from defective state (L. *vitium*) and injury (L. *trauma*). (Ref. with propos. II – 1.6.)

Diagnosis, Prognosis

In those 10 definitions of diagnosis and prognosis there is little commonality between the two as for the proximate genera of the concepts (even though the terms differ in their respective prefixes only). For diagnosis the proximate genus is most commonly given as the process of identification (of the patient's illness). For prognosis, by contrast, the proximate genus in five of the 10 definitions has to do with outcome (of a case of illness); it is given as the outcome per se or as the predicted/expected version of this.

In none of the 10 definitions is the proximate genus of either diagnosis or prognosis given as that of *knowing* (about the health of a client). But when knowing was taught to be the proximate genus of both of these concepts (and of etiognosis besides; propos. II – 1.13), no one took exception.

Good Diagnosis, Good Prognosis

Had diagnosis and prognosis been viewed as species of the genus knowing – esoteric, uncertain, about the health of a client – based on ad-hoc facts together with general medical knowledge, it would have been obvious that good diagnosis and good prognosis are characterized by the correct level of confidence/probability in the knowing, level that is warranted by the available ad-hoc facts (propos. II – 1.15, 17). ('Good prognosis' is a common misnomer for relatively favorable prognosis.)

Accuracy of a Diagnostic Test

Accuracy of a 'test' really is accuracy of its *result*; it is the *degree* to which the result generally is in accord with the truth about that which the 'test' result addresses. Degree of accord between 'test' result and the truth about the object of 'testing' is a concept that applies to *quantification* only.

A diagnostic 'test' does not address the presence/absence of a particular illness. It addresses something that bears on the probability that the illness is present; it produces a datum for incorporation into the diagnostic profile of the person (at a particular time). For example, in the pursuit of early – latent-stage – diagnosis (rule-in) about lung cancer, the result of an imaging 'test' is not positive or negative in respect to presence/absence of cancer but as to a non-calcified pulmonary nodule (as defined for the purpose); and as for the presence/absence of a non-calcified pulmonary nodule (as defined, distinct from lung cancer), the test result is either correct or incorrect, rather than characterized by a given degree of correctness/accuracy. On the other hand, degree of accuracy does characterize a 'test' (determination/measurement, quantitative) that addresses a detected nodule's size or rate of growth (which bear on the probability of the nodule's malignancy).

'Clinical epidemiologists' perpetuate misguided ideas about 'accuracy' of diagnostic 'tests,' which arguably is their favorite topic. This is much in evidence in those 10 definitions of the accuracy of a diagnostic test. An indication of how addled the ideas are is this: for the diagnosis about whichever particular illness, an arbitrarily chosen 'test' generally would produce a negative result and thus, by the prevailing definition, would have for the illness – any illness – a high 'specificity'! In reasonable terms, it is a 'test' *result* – or a diagnostic profile as a whole – that has a given degree of specificity to a particular illness, meaning that it is, to a given degree, pathognomonic – indicative of – the presence (or absence) of that illness. And a major flaw in the preoccupation with measures of a 'test's' diagnostic 'accuracy' is the continual treatment of them as though singular in value/magnitude – independent of the pre-test profile.

With rare exceptions – glucose tolerance test (for diagnosis about glucose intolerance, type 2 diabetes) is one of these, exercise test (for coronary stenosis) is another – *so-called diagnostic tests are not tests* in the general meaning of 'test': they do not represent challenges to evaluate structural integrity or functional capacity.

APPENDIX – 3. ASSIGNMENTS TO THE STUDENTS

Assignment 1

A. Identify, and explain, deficiencies of scholarly intellection manifest in the five definitions of clinical research and/or clinical epidemiology in propos. I – 3.1.

B. Comment on this: "Since its beginnings in 1929, the College has recognized the value of research in the regulations governing resident educational programs. In 1951, the College stated that a knowledge of basic sciences was necessary for a proper understanding of any specialty and encouraged a year of full-time graduate student-level training in research as well as teaching in a basic science department of a recognized medical school. ... In the [25 years since 1975], research experience as essential for all residents has been more strongly encouraged [by the College]."

Reference: The Royal College of Physicians and Surgeons of Canada. The Evolution of Specialty Medicine 1979–2005; p. 94.

C. In EBM the first 'step' (out of five) is "converting the need for information ... into an answerable question" (ref.). Comment on the tenability/untenability of this precept from the vantage of rational medicine.

Reference: Sackett DL, Straus SE, Richardson WS, et alii. Evidence-Based Medicine. How to Practice and Teach EBM. Second edition. Edinburgh: Churchill Livingstone, 2000; p. 3.

D. If 'clinical epidemiology' is relevant for all clinical residents and fellows to study, does it follow that all of their preceptors should be qualified to teach it? Similarly, given that all clinical residents and fellows are supposed to be graduates of medical schools, does it follow that all competent clinicians are qualified to teach all of the subjects in the medical curriculum that actually are relevant for the practice of whichever discipline of clinical medicine?

E. Comment on the level of clinical scholarship represented by the idea that "clinical epidemiology" is "a basic science for clinical medicine" (ref.).

Reference: Sackett DL, Haynes RB, Gyatt GH, Tugwell P. Clinical Epidemiology. A Basic Science for Clinical Medicine. Second edition. Boston: Little, Brown and Company, 1991.

O. S. Miettinen, *Up from CLINICAL EPIDEMIOLOGY & EBM*,
DOI 10.1007/978-90-481-9501-5_21, © Springer Science+Business Media B.V. 2011

Assignment 2

A. Specify the proximate genus of medicine in three different, mutually consistent ways, and comment on the relative merits of these.
B. Define the clinical concept of case.
C. Identify the respective families of concepts in which the proximate genera of the concepts of illness and diagnosis belong.
D. Comment on the expression, 'To make a diagnosis of [illness I] requires … '
E. What is generally required for genuine diagnosis to be possible?
F. Is correct diagnosis without any discriminating ad-hoc facts possible?
G. Distinguish, succinctly, between etiogenesis and pathogenesis, and between etiogenesis and etiognosis; and comment on the idea that etiology is "Literally, the science of causes" (according to A Dictionary of Epidemiology).
H. What is, and what should be, the concept of 'good prognosis'?
 I. Is prediction the proximate genus of (medical) prognostication?
J. Is there need for the terms 'diagnostication' and 'etiognostication'? What about 'etiognosing' and 'prognosing'?

Assignment 3

A. When a set of facts is known about a patient presenting with a complaint, the doctor translates this set into deeper, esoteric knowing. What questions should be addressed in this translation?
B. Regarding that set of facts, comment on the relevance of inclusion among those facts

   (i)   the particular practice within the type ('specialty') of practice;
   (ii) the type of practice; and
   (iii) the time/place of the patient's presentation (for diagnosis).

C. Regarding bullet wound,

   (i)   what might Robert Koch have taken to be its cause? and
   (ii) what actually is the proximal cause?

D. Is 'borderline hypertension' causal to stroke (in some instances)? Also, comment on whether the term 'hypertension' is apposite for the concept to which it refers.
E. Comment on the relative merits of these two outlooks in intervention-prognosis (bearing on the decision about the intervention):

   (i)   the need is to address the effects of the intervention (relative to its alternative) on the basis of knowledge about differences in the probabilities of prospective events/states;
   (ii) the need is to address future course – descriptive – conditionally on the intervention and its alternative, respectively.

F. If a person is free of any sickness representing potential manifestations of lung cancer but is worried about possibly having a detectable case of this disease and wonders about being screened for it,

   (i) would (s)he do well seeking advise from some office or official of community medicine in preference to a pulmonologist clinician?;

   (ii) would (s)he reasonably expect the pursuit of the detection (rule-in diagnosis), were (s)he to go for it, to be carried out by an epidemiologist (a community doctor) rather than (a team of) clinicians?;

   (iii) should his/her decision about the diagnostic pursuit be preordained by public policy (of community doctors)?

Assignment 4

Suppose a diagnostic domain is defined by (the presence of) symptom S, classified as either mild (S_-) or severe (S_+); and test T is performed in this domain, with its result classified as negative (T_-) or positive (T_+). Let the probabilities/prevalences of the presence of illness I be represented thus:

	T_-	T_+	
S_-	P_{00}	P_{01}	(Table 1)
S_+	P_{10}	P_{11}	

And let the distribution of the instances from the domain be represented by these probabilities:

	T_-	T_+	Total	
S_-	Q_{00}	Q_{01}	$Q_{0.}$	(Table 2)
S_+	Q_{10}	Q_{11}	$Q_{1.}$	
Total	$Q_{.0}$	$Q_{.1}$	1	

Let X_1 be indicator of S_+ (i.e., 1 if S_+, 0 otherwise), X_2 indicator of T_+, and $X_3 = X_1 X_2$.

A. Suppose the 'saturated' model, $\mathrm{logit}(P) = B_0 + \Sigma_i B_i X_i$, $i = 1, 2, 3$, is adopted.

   (i) What are the values of the parameters of the model in terms of the probabilities in Table 1?

   (ii) What are the values of the probabilities in Table 1 in terms of the parameters of this model?

   (iii) How well does the model describe the value of P as a function of S and T (within the S-based domain)? Why?

B. Suppose the 'additive' model, $\text{logit}(P) = B_0 + B_1X_1 + B_2X_2$ is adopted.

   (i) On what condition is this model fully consistent with the pattern in Table 1, accurately descriptive of it?

   (ii) What are to be the interpretations of the parameters of this model?

C. The 'sensitivity' of a test with a binary result (like T here) for illness I is said to be the probability of T_+ given I present, and 'specificity' for I the probability of T_- given I absent.

   (i) In the S-based domain at issue here, what are the values of $\Pr(T_+ | I)$ and $\Pr(T_- | \bar{I})$, respectively?

   (ii) Are these the test's 'sensitivity' and 'specificity' (for I), respectively, as defined? What is $\Pr(T_+ | I, S_+)$? Is this 'sensitivity'? Comment.

   (iii) What can be surmised to be the 'specificity,' for I, of an arbitrarily chosen test (binary in result), as an approximation? Comment.

   (iv) What is the correct pre-test probability of (i.e., correct pre-test diagnosis about) I being present in the example here?

   (v) What is the corresponding correct post-test diagnosis about I, given T_+?

   (vi) With $\text{logit}(P) = B_0 + B_1X_1$ the model for the pre-test probability, what are the values of B_0 and B_1, respectively, in the example here?

Assignment 5

Consider patients presenting for diagnosis with 'chief complaint' about symptom S, classified as either 'mild,' 'moderate,' or 'severe.' A test is performed at some time $T = t$ after the onset of the symptom, and it gives some quantitative result $Q = q$ units. No other diagnostic indicators are considered (in this first stage of the pursuit of diagnoses).

A. Specify the diagnostic domain at issue.

B. Specify – superficially – the set of diagnostic indicators accounted for (at this stage, to define subdomains).

C. Specify the nature of the scale – and thus specific essence – for each of the diagnostic indicators.

D. Translate the set of diagnostic indicators into a corresponding set of statistical variates.

E. What are some examples of alternatives to the model that involves that set of Xs (specified in part D)?

F. Is there, in this example, a fundamental difference between the pre-test, history-based indicator and the test-based indicator(s) in their treatment for modeling the post-test probability?

G. Are the diagnostic indicators in this example prone to be mutually correlated? Explain.

H. What bearing, if any, does/would the indicators' correlatedness (within the domain) have on the descriptive appropriateness of the model implied by the variates defined in part D?

Assignment 6

Recall Assignment 4. Suppose the values of the (P, Q) pair of parameters are these:

	T_-	T_+
S_-	(0.05, 0.50)	(0.60, 0.10)
S_+	(0.40, 0.10)	(0.95, 0.30)

And suppose the (additive) model $logit(P) = B_0 + B_1X_1 + B_2X_2$ is used for the form of the post-test probability function in the domain at issue (with X_1 and X_2 as in Assignment 4).

A. Is the form of this post-test model consistent with the pattern of the probabilities? Explain.
B. What are the values of the parameters of this model?
C. The corresponding pre-test model is of the form $logit(P) = B_0 + B_1X_1$.

   (i) Is this model fully valid? Explain.
   (ii) What are the values of the parameters in this model?

D. Based on the post-test model (incl. the values of its parameters), what is the range of the possible post-test probabilities when the pre-test profile is S_-?; and what is it when that profile is S_+? Explain.
E. Specify a suitable form for the model for $Pr(T_+)$.

   (i) What are the values of the parameters of this model?
   (ii) What is, per that model for $Pr(T_+)$, the probability that the post-test probability (per the model for this) will exceed 0.90, given pre-test profile S_+?

Assignment 7

A. What is the concept of 'drug interaction' in the etiogenesis of adverse effects of medication uses? Is the term apposite for its meaning?
B. With X_1 and X_2 the indicators of recent use of drug A and drug B, respectively, what is the implication of the additive log-linear modeling for the rate (incidence density) of an adverse event, modeling in which $L = B_0 + B_1X_1 + B_2X_2 +$ confounder terms? Specifically, what is implied about the magnitude of the effect of a given one of the medication uses on the magnitude of the effect of the other on the rate of the event's incidence density?

C. Consider, for the log metameter of incidence density in the context of those two medication uses, the model that includes, additionally, $X_3 = X_1 X_2$, $X_4 =$ dosage of drug A (numerical value of), $X_5 =$ dosage of drug B, and $X_6 = X_4 X_5$.

  (i) What is implied to be the form of the rate-ratio function for the use of medication A, with no use of A as the alternative.

  (ii) If the alternative for the use of medication A, at a given dose, with no use of drug B, is that use of drug B at the same level of dose, what now is the implied form of the etiognosis-relevant rate ratio?

D. For etiognosis about an adverse event that can be an (idiosyncratic) drug reaction, what particulars of the history of the medication's use should generally be accounted for in the model for the reaction's incidence density?

APPENDIX – 4. TO THE ASSIGNMENTS, THE TEACHER'S RESPONSES

Assignment 1: Teacher's Responses

A. Clinical research

Fletcher & Fletcher. A science has a cohesive 'material' – as distinct from 'formal' – object, as do, for example, cardiology and neurology (as sciences, rather than disciplines of practice). Epidemiology as a research discipline is not a science, just as morphology isn't; instead, sciences (e.g., cardiology and neurology) involve epidemiological (as well as morphological) issues as for the formal objects. Nor is "making predictions about individual patients" science, but ideally it is application of science. Diagnostic and etiognostic probability-settings obviously are not *pre*dictions. Nor actually are the prognostic counterparts of these predictions: setting a prognostic probability for something – especially if that probability is low – is not tantamount to foretelling – predicting – that something. To the extent that something (qualitative) actually is predicted in clinical medicine, the prediction is either correct or incorrect, rather than "accurate" or inaccurate (as commonly is the case with, e.g., weather forecasts/predictions). Describing gnosis-relevant clinical research as "counting clinical events in groups of similar patients and using strong scientific methods ..." is an incongruous juxtaposition of trivializing and hubris.

Hulley et alii. A given science is constituted by the research on its material object and by the body of knowledge resulting from this. There is no "science of doing clinical research." The melange that is said to constitute this has no rhyme or reason. For example, whereas clinical medicine is to be distinguished from community medicine – from epidemiology, that is (propos. II – 1.10) – it is a gross 'category error' to (clearly) imply that epidemiological research is one of the "forms" of "doing clinical research." (These misunderstandings echo those of 'clinical epidemiologists' at large.)

Glasser. There is no intellectual virtue in taking a "middle of the road" to a definition, any more than in taking it between truth and falsehood; there is to be a tenable rationale for the adopted definition; it is to be logically admissible (propos. II – 1.2). The definition actually adopted by Glasser serves as the epitome of rationality-challenged conception of the essence of clinical research.

O. S. Miettinen, *Up from CLINICAL EPIDEMIOLOGY & EBM*,
DOI 10.1007/978-90-481-9501-5_22, © Springer Science+Business Media B.V. 2011

Gauch. Implicitly, but unjustifiably, clinical research is equated with medical research. Most of medical research is 'bench' research and, as such, it is not mostly 'drug' research. Not even a semblance of definition of clinical research is given, even though the term is in the book's title.

Grobbee & Hoes. Diagnosis, etiognosis, and prognosis are not objects of clinical research; they are applications of quintessentially 'applied' clinical research (propos. II – 1.13, 15–17). Effects of interventions are not extrinsic to prognostic research; they are of central concern in it (sect. III – 4). And the common "principles and methods" duality is untenable: the book is about methodologic principles, among other issues of the theory of clinical research.

For a logically admissible, *justifiable conception* of the essence of clinical research, the proximate genus (propos. II – 1.1) is, quite obviously, medical research – for which a tenable definition is implied by proposition I – 2.4: research for the advancement of the arts of medicine. Within this genus, the specific difference (propos. II – 1.1) in the essence of clinical research is, quite obviously again, that the purpose is to advance the arts of clinical medicine (as distinct from those of community medicine, of epidemiology, i.e.; propos. II – 1.10). Thus, clinical research is (medical) *research for the advancement of the arts of clinical medicine*.

In clinical research (as just defined), a *major distinction* is that between quintessentially 'applied' and only in-essence 'applied' research (propos. I – 2.5). An eminent example of this distinction has to do with research (prognostic) for the knowledge-base of medication/drug use as distinct from research with the aim of making a new medication available for use (i.e., research in the overall effort of 'drug development') or making an already available medication to have a new indication for use. Clinicians are concerned with the knowledge derived from the former type of research, while the latter type is relevant in terms of what innovation, if any, it brings for consideration in practice – and for quintessentially 'applied' research to address.

Throughout this text the word 'applied' has been used in quotation marks, thus acknowledging it as established medical jargon (propos. I – 2.4) yet indicating reserve in its acceptance. Knowledge derived from quintessentially 'applied' clinical research is, by definition, supposed to be *applicable* but it isn't necessarily applied; and knowledge produced by merely in-essence 'applied' clinical research is only very exceptionally applied in the form of a product advancing the practice of clinical medicine.

B. Research experience in residency

It is *not true* that "knowledge of basic sciences [is] necessary for a proper understanding of any specialty," nor does one gain that knowledge in a *single* basic-science department (in one year). The first one of these mutually incoherent ideas is a misrepresentation of the Flexnerian fallacy concerning the essence of, and preparation for the practice of, scientific medicine (propos. I – 2.8). Regardless, critical for competent practice is not understanding but *knowing* (propos. II – 1.7, 13).

(The teacher of this course keeps failing to find a professor, let alone a non-academic practitioner, of internal medicine who remembers the molecular structure of aspirin or the outlines of Krebs' cycle.)

C. Asking an answerable question

In rational medicine one asks appropriate, relevant questions, answerable or not (propos. IV – 1.4).

D. Should clinical preceptors know that which all clinical residents and fellows, regardless of discipline, are supposed to learn?

To be competent in their role, clinical preceptors should, of course, master whatever their clinical residents and fellows are reasonably supposed to learn, including about 'clinical epidemiology' (propos. I – 2.13). By the same token, all competent clinicians, regardless of discipline, actually are qualified to teach all of the subjects that justifiably belong in the curricula of medical schools, which clinical residents and fellows, regardless of discipline, truly should have learned (propos. I – 2.14). (A competent clinician typically retains very little of that which once was taught to him/her in the undifferentiated medical school, such as it still is.)

E. 'Clinical epidemiology' as a basic science

Medical sciences are customarily classified as 'basic' or 'applied' (propos. I – 2.4). If epidemiology were a science (cf. '1.A' above) and if there were a clinical version of it, then clinical epidemiology (would exist and) obviously would be one of the 'applied' medical sciences, not a basic science. Both of the premises are false, however: 'clinical epidemiology,' insofar as such an entity is regarded as being real (ontologically admissible), is not an 'applied' medical science, much less a basic medical science. It actually defies rational definition, even (cf. propos. I – 5.2).

Assignment 2: Teacher's Responses

A. Proximate genus of medicine

Three possibilities (consistent with proposition II – 1.7):

(1) professional pursuit of esoteric ad-hoc knowing, and teaching accordingly (propos. II – 1.7);
(2) art of gaining esoteric ad-hoc knowing, and teaching accordingly; and
(3) art of supplying esoteric answers.

Number 1 is to be preferred. For, even doctors themselves do not necessarily know the scholarly concept of art.

B. The clinical concept of case

Sometimes, improperly, a person (patient) is said to be a case. In proper terms, case is an instance of something; for example, a patient may have (rather than be) a case of pneumonia.

C. Illness and diagnosis: proximate genera

Illness has somatic anomaly as its proximate genus, diagnosis has knowing as its proximate genus.

D. Making a diagnosis

To 'make a diagnosis of ...' is a very poor – though very common – way to refer to the *pursuit* and *attainment* of diagnosis *about* ... One does not 'make' knowing (which diagnosis is).

E. Requirement for diagnosis

Diagnosis is ad-hoc, particularistic knowing on a level deeper than the available facts (note 'esoteric' in 2.A above). It requires general (abstract) medical knowledge as an added input (propos. II – 1.14).

F. Diagnosis without facts

Perfectly possible, in principle. The need is to know about the prevalence of the illness in the domain of diagnosis at large, without specification of subdomain (on the basis of indicators of risk for, or manifestations of, the illness).

G. Etiogenesis vis-à-vis pathogenesis

A case of illness comes into being through the process of its pathogenesis, meaning the sequence of changes from normal tissue to the illness-definitional anomaly; the etiogenesis of the case is the (sequence of) causal influence(s) that initiated and/or advanced the pathogenesis. Etiognosis is knowing about the etiogenesis of a case of illness. Etiology is no more literally the science of causes than tautology is literally the science of unnecessary repetition.

H. Good prognosis

The common concept of 'good prognosis' now is one of the possible properties of an illness – that it generally has a relatively favorable course. But insofar as prognosis is understood to have knowing as its proximate genus (cf. 2.C), good prognosis must mean more-or-less-correct prognosis – justifiable as to the probability in it – however unfavorable it might be.

I. Prediction vis-à-vis prognostication

Prognostication (in clinical medicine) is knowing about the probability (relative frequency) of a prospective phenomenon of health in a person; predicting a phenomenon of health is declaring/forecasting that it will occur. Thus, prognostication is not (limited to) prediction.

J. Gnostication vis-à-vis gnosing

Given the established term 'prognostication,' 'diagnostication' and 'etiognostication' should – by analogy – be regarded as legitimate. Similarly, given the established term 'diagnosing,' 'etiognosing' and 'prognosing' should be regarded as legitimate.

Assignment 3: Teacher's Responses

A. Translating diagnostic profile into diagnosis

The questions that need to be addressed (and, indeed, answered) are these: What is the full set of possible underlying illnesses (explanatory of the manifestational profile); and, What are the respective diagnoses about – probabilities of – these?

B. Particularistic elements in a diagnostic profile

Diagnosis is achieved by bringing general (abstract) medical knowledge to bear on the available particularistic facts. For this to be the case, those particularistic facts must pertain to general things (about which there can be general knowledge).

Nothing about a practice qualifies as an entry among the facts, nor does the time or place of the presentation.

C. Cause of bullet wound

Just as Koch took *M. tuberculosis* to be the universal cause of tuberculosis, he presumably would have taken bullet to be the universal cause of bullet wound. But just as the universal (necessary) cause of tuberculosis actually is effective exposure to the mycobacterium in conjunction with susceptibility to its invasion (propos. II – 1.28), so the universal cause of bullet wound is exposure to the trajectory of effectively fast motion of a bullet (to which susceptibility is universal).

D. Hypertension as a cause of stroke

A question about a given level of hypertension as a possible cause of stroke is meaningless without specification of the alternative (propos. II – 1.25). Relative to severe, malignant hypertension, borderline hypertension is preventive of stroke. The term 'hypertension' is less than apposite for high pressure (rather than high tension); and rather than high pressure, even, at issue actually is high peripheral vascular resistance (in systemic, as distinct from pulmonary, 'hypertension').

E. Intervention as a topic in prognosis

To take a simple example, for a decision about an intervention it is more meaningful to know the probability of fatal outcome with the contemplated intervention and with its alternative than to know merely the difference between these two intervention-conditional prognoses (cf. propos. II – 1.31, III – 4.19).

F. Advise about screening for lung cancer

Absurd though it is, epidemiologists concerned with community oncology presume to be the experts on screening, for lung cancer or whatever illness. (They think of screening as a single test, and its application as a matter of community intervention – and a preventive one at that – rather than community diagnosis!) This they presume even though the pursuit of the cancer's early, latent-stage diagnosis is pursued by a team of professionals in clinical medicine, and the treatment of the diagnosed case also is a clinical matter. Decision about screening for lung cancer, as about anything else in clinical medicine, is the province of the doctor's client. Epidemiologists should not interfere with this by their very ill-justified public policies.

Assignment 4: Teacher's Responses

Diagnosis with two binary indicators, concerning the severity of symptom (S) and the result of a test (T).

Diagnostic probabilities

	T_-	T_+
S_-	P_{00}	P_{01}
S_+	P_{10}	P_{11}

Patients' distribution

	T_-	T_+	Total
S_-	Q_{00}	Q_{01}	Q_0
S_+	Q_{10}	Q_{11}	$Q_1.$
Total	$Q_{.0}$	$Q_{.1}$	1

X_1: indicator of S_+; X_2: indicator of T_+; $X_3 = X_1X_2$.

A. Model: $\text{logit}(P) = B_0 + \Sigma_i B_i X_i, i = 1, 2, 3$.

   (i) $B_0 = \text{logit}(P_{00})$; $B_1 = \text{logit}(P_{10}) - \text{logit}(P_{00})$; $B_2 = \text{logit}(P_{01}) - \text{logit}(P_{00})$; $B_3 = \text{logit}(P_{11}) - (B_0 + B_1 + B_2)$.

   (ii) $P_{00} = 1/[1 + \exp(-B_0)]$; $P_{10} = 1/[1 + \exp(-B_0 - B_1)]$; $P_{01} = 1/[1 + \exp(-B_0 - B_2)]$; $P_{11} = 1/[1 + \exp(-B_0 - B_1 - B_2 - B_3)]$.

   (iii) The model describes the probability pattern perfectly. For, the model has as many parameters as there are distinct probabilities to specify; no particular pattern of the probabilities is required for such a 'saturated' model to be perfectly descriptive of the pattern.

B. Model: $\text{logit}(P) = B_0 + \Sigma_i B X_i, i = 1, 2$.

   (i) It must be that, with the saturated model, $B_3 = 0$; that is, that $\text{logit}(P_{11}) = \text{logit}(P_{00}) + [\text{logit}(P_{10}) - \text{logit}(P_{00})] + [\text{logit}(P_{01}) - \text{logit}(P_{00})]$; that is, that additivity of the logits obtains.

   (ii) $B_0 = \text{logit}(P_{00})$; $B_1 = \text{logit}(P_{10}) - \text{logit}(P_{00}) = \text{logit}(P_{11}) - \text{logit}(P_{01})$; $B_2 = \text{logit}(P_{01}) - \text{logit}(P_{00}) = \text{logit}(P_{11}) - \text{logit}(P_{10})$.

C. 'Sensitivity' and 'specificity' here

   (i) $\text{Pr}(T_+ | I) = (Q_{01}P_{01} + Q_{11}P_{11})/(Q_{00}P_{00} + Q_{10}P_{10} + Q_{01}P_{01} + Q_{11}P_{11})$.
$\text{Pr}(T_- | \bar{I}) = [Q_{00}(1 - P_{00}) + Q_{10}(1 - P_{10})]/[Q_{00}(1 - P_{00}) + Q_{10}(1 - P_{10}) + Q_{01}(1 - P_{01}) + Q_{11}(1 - P_{11})]$.

   (ii) Those are the test's 'sensitivity' and 'specificity' for I, per the respective definitions.
$\text{Pr}(T_+ | I, S_+) = Q_{11}P_{11}/(Q_{10}P_{10} + Q_{11}P_{11})$. If 'sensitivity' indeed were relevant, this – together with $\text{Pr}(T_+ | I, S_-)$ – would be more so, given that the S status is known (before the test). But the genuine interest is in $\text{Pr}(I | S_-, T_-) = P_{00}$, etc; and it also is in $\text{Pr}(T_+ | S_+)$ and $\text{Pr}(T_+ | S_-)$ (propos. III – 2.4).

(iii) An arbitrarily chosen test with result classified as either positive or negative presumably would give a negative result in some 95% of instances in which I is absent, meaning that it is highly 'specific' to I! Any diagnostic test is highly 'specific' to whatever illness! Such is the disarray of concepts (and terms) in the favorite topic of 'clinical epidemiologists' – the 'accuracy' of diagnostic tests, that is.

(iv) Pre-test probability of I: If S_-, then $(Q_{00}P_{00} + Q_{01}P_{01})/Q_0$; if S_+, then $(Q_{10}P_{10} + Q_{11}P_{11})/Q_1$.

(v) Post-test probability, like the pre-test probability, depends on the pre-test profile (as to S). If S_-, then $\Pr(I\,|\,S_-,\,T_+) = P_{01}$; if S_+, then $\Pr(I\,|\,S_+,T_+) = P_{11}$.

(vi) Under $\mathrm{logit}(P) = B_0 + B_1X_1$ (for pre-test probability), $B_0 = \mathrm{logit}[(Q_{00}P_{00} + Q_{01}P_{01})/Q_{0\cdot}]; B_1 = \mathrm{logit}[(Q_{10}P_{10} + Q_{11}P_{11})/Q_{1\cdot}] - B_0$.

Assignment 5: Teacher's Responses

A. The diagnostic domain here is simply that of symptom S as the 'chief complaint.'

B. Two diagnostic indicators: severity of S and result of test Q at time T.

C. The symptom is specified on an ordinal scale of severity, the categories (ordinal) of the severity scale being 'mild,' 'moderate,' and 'severe.' The test-based indicator is two-dimensional: temporal and result-specifying, both of these quantitative.

D. One possibility: symptom-based variates X_1 = indicator of 'moderate' (X_1 = 1 if 'moderate,' 0 otherwise); X_2 = indicator of 'severe'; test-based variates $X_3 = T$, $X_4 = Q$, $X_5 = X_3X_4$ (numerical values of T and Q).

E. Based on X_1 through X_4, the model could additionally involve, for example, $X_6 = X_3^2$ and $X_7 = X_4^{1/2}$. Or without any addition, one equivalent of the model in part E would involve X_2 = indicator of 'mild' (instead of 'severe').

F. The pre-test vs. post-test distinction is simply a matter of whether the test-based variates (X_3 and X_4) are involved in the model. Nothing fundamental in this (cf. propos. III – 2.2).

G. Severity of symptom(s) and level(s) of test result(s) have a propensity to be correlated – positively, as both tend to reflect the severity of the illness.

H. That correlatedness has no bearing on the appropriateness of the regression models. The models here address diagnostic probabilities conditional on X_1 through X_4, and these probabilities do not depend on how the cases from the domain are distributed by X_1 through X_4, including mutual correlations of these. (This is in sharp contrast to transitions from pre-test probabilities to post-test probabilities by means of likelihood ratios that are not specific to the pre-test profiles involved.)

Assignment 6: Teacher's Responses

In Assignment 4 the values of (P, Q) could have been these:

	T_-	T_+
S_-	$(0.05, 0.50)$	$(0.60, 0.10)$
S_+	$(0.40, 0.10)$	$(0.95, 0.30)$

The post-test model might be the additive one: $\text{logit}(P) = B_0 + B_1X_1 + B_2X_2$, with X_1 indicator of S_+ and X_2 indicator of T_+.

A. The $\text{logit}(P)$ differences in the two rows are $\log(0.60/0.40) - \log(0.05/0.95) = 3.35$ and $\log(0.95/0.05) - \log(0.40/0.60) = 3.35$; and in the two columns they also are identical, 2.54 each. Thus, the additive model for $\text{logit}(P)$ is accurately descriptive of the pattern of probabilities.

B. $B_0 = \text{logit}(0.05) = -2.94$; $B_1 = \text{logit}(0.40) - \text{logit}(0.05) = \text{logit}(0.95) - \text{logit}(0.60) = 2.54$ (cf. part A above); $B_2 = \text{logit}(0.60) - \text{logit}(0.05) = \text{logit}(0.40) = 3.35$ (cf. part A above).

C. The pre-test model (implied by the post-test model) is $\text{logit}(P) = B_0 + B_1X_1$. Given that X_1 is binary, the two probabilities are fully described by this two-parameter model. $B_0 = (0.50 \times 0.05 + 0.10 \times 0.60)/(0.50 + 0.10) = 0.14$; $B_1 = (0.10 \times 0.40 + 0.30 \times 0.95)/(0.10 + 0.30) - B_0 = 0.57$.

D. The post-test model is $\text{logit}(P) = -2.94 + 2.54\,X_1 + 3.35\,X_2$. Given S_- (i.e., $X_1 = 0$), $\text{logit}(P) = -2.94 + 3.35\,X_2$, and the corresponding possibilities for the post-test $\text{logit}(P)$ are -2.94 and $(-2.94 + 3.25) = 0.41$, which respectively translate into $1/[1 + \exp(2.94)] = 0.05$ and $1/[1 + \exp(-0.41)] = 0.60$ for P. Given S_+ (i.e., $X_1 = 1$), $\text{logit}(P) = -2.94 + 2.54 + 3.25\,X_2$, implying for P the possible values 0.40 and 0.95. (These results accord with the table above.)

E. $\Pr(T_+) = 1/[1 + \exp(-B_0 - B_1X_1)]$, with X_1 indicator of S_+ (as before). $B_0 = \text{logit}[0.10/(0.50 + 0.10)] = -1.61$. $B_1 = \text{logit}[0.30/(0.10 + 0.30)] - (-1.61) = 2.71$. $\Pr(I\,|S_+, T_+) = 0.95 > 0.90$ (cf. part D above). $\Pr(T_+\,|S_+) = \Pr(T_+\,|X_1 = 1) = 1/[1 + \exp(1.61 - 2.71)] = 0.75$.

Assignment 7: Teacher's Responses

A. 'Drug interaction' in the etiogenesis of adverse events means that the probability with which a given drug was causal to the event depends on whether the other drug was used. The meaning is that the effect of one drug's use depends on – is modified by – the other drug's use; it is not that the drugs interact, the molecules influencing each other. Misnomer (cf. 'gene-environment interaction.')

B. The model with $L = B_0 + B_1X_1 + B_2X_2$, with X_1 and X_2 the indicators for the use of drug A and drug B, respectively, generally would be for the logarithm of the event's incidence density (ID), implying additivity of effects on log(ID) and, hence, multiplicativeness of them (in ratio terms) on ID itself.

C. An expanded model might indeed involve $X_3 = X_1X_2, X_4 =$ dosage of A, $X_5 =$ dosage of B, and $X_6 = X_4X_5$. The implication now is that for A $(X_1 = 1)$ vs. $\bar{A}$ $(X_1 = 0)$ the rate ratio is

$$IDR = \exp(B_0 + B_1 + B_2X_2 + B_3X_3 + B_4X_4 + B_5X_5 + B_6X_4X_5)/$$
$$\exp(B_0 + B_2X_2 + B_5X_5)$$
$$= \exp(B_1 + B_3X_3 + B_4X_4 + B_6X_4X_5).$$

If the alternative to A without B is no A but B at the same level of dose $(X_5 = X_4)$, then

$$IDR = \exp(B_0 + B_1 + B_4X_4)/\exp(B_0 + B_2 + B_5X_4)$$
$$= \exp[(B_1 - B_2) + (B_4 - B_5)X_4].$$

D. The need generally is to consider 'recent' use as the etiogenetically relevant period (where the causation could have taken place). Earlier use matters in that it serves to weed out those susceptible to the reaction to the drug from among recent users and it also tends to have the corresponding selectivity consequence differentially between recent users and nonusers in respect to susceptibility to the reaction to other causes of the adverse event (Miettinen OS, Caro JJ. J Clin Epidemiol 1989; 42: 325–31).

APPENDIX – 5. MORE ON GARNERING EXPERTS' TACIT KNOWLEDGE

Given that early treatment of an acute coronary event – unstable angina or myocardial infarction – has developed so that it now commonly serves to avert the episode's otherwise fatal outcome, swift and expert diagnosis of *acute myocardial ischemia* – practical rule-in or rule-out diagnosis of AMI – as the basis for ER (emergency-room) decision about referral to the CCU (coronary care unit) now deserves to be viewed as an eminent feature of high-quality healthcare.

The teacher of this course, in collaboration with Steurer and others in the Horten Center for Practice-oriented Research and Knowledge Transfer of the University of Zurich, Zurich, Switzerland, has been working on a (demonstration) project on garnering experts' tacit knowledge concerning this diagnosis. The colleagues constituting the expert panels were experienced *ER doctors* in hospitals in Switzerland, identified by professors of internal medicine in the country.

The *domain* of this diagnosis – and hence of the diagnostic probability function being developed – was taken to be the patient (of either gender), 30 years of age or older, who presents with the chief complaint of very recent – within 12 hours – episode of *acute dyspnea and/or acute chest 'pain'* (retrosternal) sustained at rest.

Relevant to the specification of the diagnostic indicators, the *context* of this diagnosis was taken to be presentation in an ER with an associated CCU, specifically *arrival at the ER*, though with some delay possibly occasioned by the need to wait for the results of the enzyme tests (in the absence of pathognomonic signs of AMI in the ECG).

For this context, we defined a total of 41 diagnostic indicators. The first six of these have to do with the particulars of the chief complaint (dyspnea and/or chest 'pain,' while the associated other symptoms and signs are addressed separately). These six indicators were:

(1) time since the onset of the symptom(s) (until arrival at the ER): number of hours;
(2) duration of the symptom(s): minutes/hours;
(3) dyspnea: no/yes;
(4) type of chest 'pain': sharp / burning / pressure or tightness / no chest 'pain';
(5) aggravation of chest 'pain' by inspiration or change of position: no / yes / no chest 'pain'; and
(6) radiation of chest 'pain' to the left shoulder, arm and/or neck / chin: no / yes / no chest 'pain.'

O. S. Miettinen, *Up from CLINICAL EPIDEMIOLOGY & EBM*, 167
DOI 10.1007/978-90-481-9501-5_23, © Springer Science+Business Media B.V. 2011

Somewhat arbitrarily, 48 was chosen as the number of profiles (all different, based on the 41 indicators) presented to the members of the panel of experts.

For the specification of the 48 profiles, the point of departure was a *perfectly orthogonal 48 × 41 matrix of (0, 1) values*, 24 of each value in each of the 41 columns. In this matrix the first 47 (of 48) values in the first column (corresponding to the first indicator, above) was the sequence (from profile 1 to profile 47): 1, 1, 1, 1; 1, 0, 1, 1; 1, 1, 0, 0; 1, 0, 1, 0; 1, 1, 1, 0; 0, 1, 0, 0; 1, 1, 0, 1; 1, 0, 0, 0; 1, 0, 1, 0; 1, 1, 0, 0; 0, 0, 1, 0; 0, 0, 0. The second column of 47 values (in this 47 × 41 matrix) was obtained by shifting the first column 'cyclically' down by one place/row: 0, 1, 1, 1; 1, 1, 0, 1; 1, 1, 1, 0; 0, 1, 0, 1; 0, 1, 1, 1; 0, 0, 1, 0; 0, 1, 1, 0; 1, 1, 0, 0; 0, 1, 0, 1; 0, 1, 1, 0; 0, 0, 0, 1; 0, 0, 0. The third column was derived analogously from the second column; etc. Finally, row 48 was added: all zeros. (Ref.: Plackett RL, Burman JP. Biometrika 1946; 33: 305–25.)

While this 48 × 41 matrix represents, remarkably, a perfectly 'factorial' design even though $2^{41} = 2.2 \times 10^{12}$, the diagnostic indicators of concern here do not all have binary scales; and even those that are binary would not generally be well represented by the two-point design (of equal allocation to the two categories). Type of chest 'pain' serves as an example: Where applicable (i.e., given chest 'pain'), the scale of the type of chest 'pain' is trichotomous (cf. above); and in these instances the three-point design isn't desirable either. Very few instances of 'sharp' as the description of the type of 'pain' will suffice to make the point that experts regard this as negatively pathognomonic for AMI (i.e., as serving to rule out AMI). 'Pain' described as 'burning' deserves more extensive attention; but in the main the concern is with 'pressure' or 'tightness' as the description of the 'pain' because of the commonality of this as the particular type of the presentation.

Among the six indicators specified above, we took that *basic matrix* to be, as such, adequate for the distribution of dyspnea (indicator #3 above) only, with '0' and '1' representing, respectively, 'no' and 'yes' (in terms of a two-point design).

Chest 'pain' was to have been present whenever 'dyspnea' had been absent, given the definition of the domain of diagnosis (above); and we elected chest 'pain' to have been present, as well, in 12 of the 24 instances in which dyspnea had been present. While the dyspnea column per se thus implied the presence of chest 'pain' in 24 of the 48 instances/rows, in the remaining 24 instances chest 'pain' was set as absent or present according as column #4 in the basic matrix had the entry '0' or '1.' This defined the entries in a column for an *auxiliary diagnostic indicator*, CP, addressing absence/presence of chest 'pain.'

This CP indicator we used in specifying, for a start, the type of chest 'pain' entries in the design matrix (in respect to the four possibilities specified above). Where CP was absent, a code ('9') denoting this was entered in the type of chest 'pain' column. As for the 36 instances with CP present, we identified each successive set of {0, 0, 0, 0, 0, 0} in column #4 of the basic matrix, going down the 36 rows with CP present; and each of these we changed into {0, 1, 1, 2, 2, 2} for the design matrix, with '0,' '1,' and '2' taken to be the codes for 'sharp,' 'burning,' and 'pressure or tightness,' respectively. Correspondingly, each of the successive sets of {1, 1, 1, 1, 1, 1} was changed to {2, 2, 2, 1, 1, 0}. The consequence was that, among the 36 instances of

CP, the numbers with those three types of CP got to be 6, 12, and 18, respectively. For aggravation of chest 'pain' by inspiration or change of position among the 36, each of the successive sequences of {0, 0, 0, 0, 0, 0} in the basic matrix, now in its column #5, were translated into {0, 0, 0, 1, 0, 0}, with '0' and '1' the codes for 'no' and 'yes,' respectively, so that only in six of the 36 instances was there (what may be a negatively pathognomonic) history of such aggravation of the 'pain,' Regarding radiation of chest 'pain' among the 36, the entries in column #6 of the basic matrix were used without any changes as codes for the radiation, with '0' and '1' coding for 'no' and 'yes,' respectively. The numbers of 'no' and 'yes' thus got to be 17 and 19, respectively.

For the distribution of time since the onset of symptom(s) we opted for equal allocation into 1, 3, and 9 hours. To this end, we translated each successive set {0, 0, 0, 0, 0, 0} and {1, 1, 1, 1, 1, 1} in column #1 of the basic matrix into {1, 3, 9, 9, 3, 1} in the design matrix, with the latter numbers representing the number of hours.

For the duration of the symptom(s) we elected the design matrix to involve the values 10, 60, and 180 minutes. When the time since the onset of the symptom(s) was 1 hour (according to column 1 of the design matrix), we translated the corresponding successive sets of {0, 0, 0, 0} and {1, 1, 1, 1} in column #2 of the basic matrix into {0, 1, 1, 0}, with '0' and '1' the codes of 10 and 60 minutes, respectively, in the design matrix. When the duration was not 1 hour (but 3 or 6 hours), each successive set of {0, 0, 0, 0, 0, 0} and {1, 1, 1, 1, 1, 1} in column #2 of the basic matrix was translated into {0, 1, 2, 2, 1, 0}, with '0,' '1,' and '2' the codes for 10, 60, and 180 minutes, respectively, in the design matrix.

The distributions of the remaining 35 indicators in the design matrix were set in accordance with the principles that were adopted in the context of the first six. Different from the basic matrix, the resulting design matrix did not, of course, get to be perfectly orthogonal/factorial; but major collinearities did get to be avoided.

The members of the expert panel – expected to number three dozen – were each presented with 16 of the 48 profiles (for setting the probability of AMI) in the first go-around, and later with another 16 of the 48 – so that each of the 48 profiles was addressed by two dozen experts.

After this first phase, another set of 32 different vignettes were added to the set of hypothetical cases for the same panel of experts to address.

Once projects like this have sufficiently demonstrated feasibility and productivity, programs of improvements in Information-Age medicine can begin. Leaders within various disciplines of clinical medicine will be able to initiate and otherwise take charge of these developments. And once these efforts have come to fruition to whatever extent, to that extent practitioners of clinical medicine can have the satisfaction of ultimate professionalism: functioning on a level of quality and efficiency that is not inferior to that of any colleague (in the same discipline of clinical medicine). On the level of entire systems of healthcare, consequently, *quality assurance* and *cost containment* would inherently be well served (propos. II – 3.2); and as a side effect, even *medical academia* might undergo the needed, major improvements (propos. I – 2.13–15).

APPENDIX – 6. AN INDUSTRIAL PERSPECTIVE

K.S. Miettinen

I am honored to provide encouragement from the wider industrial perspective to the twin programs outlined here for the healthcare industry – codifying and advancing the knowledge-base of medicine and thereby advancing medical professionalism. The industrial history of parallel development of knowledge and professionalism provides a rich body of lessons, reluctantly learned, and also a story of continuous pursuit and attainment of improvement. In what follows I outline the wealth of material that future leaders of medicine now have available to them in the established practices of industry at large with respect to knowledge development and professionalism.

At the heart of the program outlined in the course is the structuring of the knowledge-base of medicine and the medical profession for efficiency and other improvement, increasingly through directly practice-serving research. Such structures are common in industry at large and have a three-tiered form, in fields as diverse as systems engineering, organizational leadership, military planning, and intelligence (information fusion). The three-tiered structure (e.g. of executive, managerial, and supervisory leadership; and strategic, operational, and tactical planning) is anchored in the center layer, which aspires to be objective. It is the development of this center layer in medicine, of the knowledge-based domain of professional practice of medicine, to which the program here points the way.

The center layer in a structure of continuous improvement represents translation of knowledge into repeatable impersonal operations, as problems in this layer should be solved in the manner of the best practitioners (professionals), in the same way every time, even though the problem in any given instance may be unfamiliar to the practitioner. This repeatability arises from shared professional knowledge applied to the structure inherent in problems at this level, and professionals converge on solutions that are correct (objective). Higher-level problems generally have no objectively correct solutions but remain personality-dependent in these, while lower-level problems typically have multiple correct solutions and are dependent on skill, which often includes elements of personal preference and style.

Progress in medicine (and other fields) comes from expansion of the scope of professionalism, properly understood, of the center layer through development of codified and disseminated knowledge to establish objective answers where

O. S. Miettinen, *Up from CLINICAL EPIDEMIOLOGY & EBM*,
DOI 10.1007/978-90-481-9501-5_24, © Springer Science+Business Media B.V. 2011

subjectivism previously reigned, while simultaneously passing responsibilities to the skill layer through development of new techniques and technology for effecting results through procedures.

Wisdom, professionalism, and proficiency are all necessary values in organizations for continuous improvement (i.e., they are always to be preferred, *ceteris paribus*), but for the center layer it is professionalism, alone, which is virtue (i.e., professionalism is always to be preferred unconditionally, and not merely *ceteris paribus*). Wisdom is the corresponding virtue at the higher level, while proficiency is the corresponding virtue at the skill level. A medical professional must subordinate his wisdom to his professionalism, since wisdom pertaining to medicine is the prerogative of society and its executive agencies (in both public and private sectors). This subordination is the obverse of the vigilance with which the medical professions defend their prerogative to define the contents of the codified, objective knowledge-base of medicine.

Medical knowledge for use by professionals in the practice of medicine is in this course described as having the form of actuarial models (specifically logistic regression models) representing probability judgments of experts facing synthetic gnostic problems (subjects existing only as gnostic profiles). This combines in one application the two broadest categories of modeling used in industry, namely physical modeling and actuarial modeling. Physical modeling is used to reach exact solutions to idealized problems (e.g. computerized prediction of strain under stress for struts based on specifications of an ideal strut), while actuarial modeling is usually used to approximately represent real observations (e.g. fitting stress/strain curves to data collected on actual struts stretched and compressed under experimental conditions). The synthetic subjects considered here are idealized problems such as those used in physical models, while the gnostic probability functions (GPFs with expert content) are actuarial models.

Among the weaknesses of actuarial modeling in industry at large and relevant for the knowledge-base of medicine is that while relationships discovered in a state of natural variation (in the absence of assignable causes of variation) may be stable, assurance of a state of purely natural variation is difficult, especially when dealing with human variation. This problem is general to the knowledge-base of medicine, but there is no better way to take action on a rational basis than to proceed despite residual doubt of knowledge after mitigating instability to the extent practical. Although strict standards for universalizing inferences from expert experience and research evidence cannot be met, there still is operational value in codifying typical expert opinions as though they represented stable knowledge, provided due care is taken to codify these opinions in a modularized and rapidly updateable way, to minimize the problems associated with technically invalid universalization. Typical expert opinion is preferable to an individual practitioner's personal opinion in any purportedly knowledge-based profession.

The designs of modern expert systems, whether rule-based reasoners or Bayesian networks or fuzzy logic systems, are generally chosen to favor speed of updating estimates of parameter values and introducing new dependencies based on new evidence, and are especially chosen to favor speed to market with a first version. This is evident in designs that systematically omit interdependencies in their models, so that complex relationships are reduced to approximations built exclusively from binary relationships, with resulting ease in pair-wise updating of estimates as well as in adding new relationships when previously unknown associations are discovered. These methods are very good for demonstrations of promise or potential, but since the answers returned by any such a system cannot be correct except by coincidence (for given the omission of complex dependencies, they cannot be systematically correct) these systems will not be trusted even if they are typically correct enough to be useful. This mistrust is due to a shortcoming of their architecture, in their form of knowledge representation.

An expert system must be reasonably correct in its content and trusted as such to be beneficial; one without the other will not do. Therefore, an expert system must have the more general structure of gnostic probability functions (GPFs), or something functionally equivalent to them, while also modularized and subject to being rapidly updated to reduce the pareto-universes over which technically invalid universalizations extend.

This, then, is the practical choice faced in codifying medical knowledge for deployment by practitioners in the field: the systems that are developed quickly by small teams of researchers and are kept up-to-date with information gleaned from published research but have a structural limit to the quality of the answers that they can give, with resulting loss of trust, versus systems that avoid the structural defect in knowledge representation (by deploying full-dimensional GPFs) but require community-wide effort to supply them with information on parameters to distinguish among subdomains, as well as rapid updating with new evidence and knowledge.

Building and maintaining medical expert systems that are be both correct and trusted is probably an undertaking too large for a single medical institution; it likely will have to be directed by government in much the same way as military research and weapons development is directed by government. Furthermore, a substantial segment of medical research will have to be structured with improvement of expert systems kept in mind from the outset. The program envisioned in this course must be coordinated on a profession-wide scale for it to succeed.

Once the commitment is made to codify practice-relevant medical knowledge and restructure medical practice on a comprehensive scale, so that practice is guided by expert systems and much of clinical research is directed toward GPFs, other changes will also inevitably become necessary. These must include abandoning the current academic practices of published papers as products and mere peer review as quality control, in favor of the rigorous and professional approaches more common in industry at large. This will mean many fewer research projects, much larger project staffs, a series of intrusive reviews – including requirements reviews, preliminary reviews, and critical review of research – as well as verification and validation of results, and acceptance testing of GPF modifications, anonymous publication of results (likely not in any journal but distributed freely) attributed only to the team's institution of affiliation, and other wrenching cultural changes sure to be resisted for a generation by leaders accustomed to the current, false paradigm.

INDEX

A
Academics, 7, 9, 14, 56, 83
Applied research, 6, 25, 158

B
Basic research, 6

C
Clinical epidemiology, 13–16, 137–138
Clinical medicine, *see* Theory of clinical
  medicine
Clinical research, *see* Theory of clinical
  research
Clinical trials, 11, 41, 74–78, 98–99, 131
CONSORT statement, 76–77, 79
Cost containment, 4, 41, 44, 57, 131, 169

D
Diagnostic research, 81, 85–86, 93–99,
  126–127

E
Education, 7, 9, 17, 25–26, 57, 151
Efficiency, 41–42, 46–49, 51, 54, 57, 63, 75,
  77, 84, 93, 131, 169, 171
Evidence-based medicine, 4, 17, 19, 45,
  56, 137
Expert systems, 4, 16, 41, 44, 56–57, 121, 131,
  133–135, 172–173

F
Flexner report, 138

G
Guidelines, 19, 75–79

I
Improvement, 131–135
Information technology, 44, 131

K
Knowledge-base, 23–31, 33–40, 41–49

L
Leadership, 3–4, 13–14, 76, 171

P
Practitioners, 5–6, 9, 14, 18–19, 42, 45, 84,
  118, 159, 169, 171–173
Professionalism, 8–9, 14, 19, 42–43, 84, 138,
  169, 171–172
Prognostic research, 49, 86–87, 108–119, 127,
  158

Q
Quality assurance, 4, 41, 44, 57, 131, 169

R
Reporting, 75–79, 102–103, 114

S
Scientific medicine, 4, 6–9, 14, 18, 56,
  133–134, 158

T
Theory of clinical medicine, 21, 23–31,
  33–40
Theory of clinical research, 76, 134, 158
Training, 9, 17, 151